Women in the History of Philosophy and Sciences

Volume 7

As the historical records prove, women have long been creating original contributions to philosophy. We have valuable writings from female philosophers from Antiquity and the Middle Ages, and a continuous tradition from the Renaissance to today. The history of women philosophers thus stretches back as far as the history of philosophy itself. The presence as well as the absence of women philosophers throughout the course of history parallels the history of philosophy as a whole.

Edith Stein, Hannah Arendt and Simone de Beauvoir, the most famous representatives of this tradition in the twentieth century, did not appear form nowhere. They stand, so to speak, on the shoulders of the female titans who came before them.

The series Women Philosophers and Scientists published by Springer will be of interest not only to the international philosophy community, but also for scholars in history of science and mathematics, the history of ideas, and in women's studies.

More information about this series at https://www.springer.com/series/15896

Jos Uffink · Giovanni Valente ·
Charlotte Werndl · Lena Zuchowski
Editors

The Legacy of Tatjana Afanassjewa

Philosophical Insights from the Work of an Original Physicist and Mathematician

Springer

Preface

This volume would not have come about without the generous support of many colleagues and friends. Firstly, we are particularly grateful to Ruth Hagengruber for the opportunity to publish this volume in the Women in the History and Philosophy of Science series. We would also like to thank all the authors who have contributed to this volume, many of which also acted as reviewers for other contributions. Harvey Brown and Michel Janssen kindly provided additional peer-reviews.

Jos Uffink would like to express a very special thanks to Tamara van Bommel, (Afanassjewa's granddaughter) for allowing me to browse through a box of Afanassjawa's papers and letters, and to Margriet van der Heijden for many helpful discussions and her assistance on this project. He also thanks the Vossius Centre for the History of Humanities and Sciences at the University of Amsterdam and the Descartes Centre for the History and Philosophy of the Sciences and the Humanities at Utrecht University, the University of Geneva, the Polytechnical University of Milano, The University of Salzburg, and the Erwin Schrödinger Institute at the University of Vienna, for financial support and hospitality, and audiences at these various locations for their feedback. He also is grateful to the University of Minnesota for a sabbatical leave, during which much of this project took shape.

Giovanni Valente would like to thank the FWF der Wissenschaftsfonds for supporting him under the Lise Meitner Programme, as well as the Department of Philosophy at University of Salzburg for hosting him during the development of this project.

Charlotte Werndl would like to thank Jeffrey Barrett, Harvey Brown, Samuel Fletcher, Laurenz Hudetz and the other contributors to this volume for valuable discussions.

Lena Zuchowski would like to thank Cornelia Zuchowski-Gemmeke and Ruth Rustemeyer for taking an early interest in this project and for suggesting the publication venue.

Minneapolis, USA — Jos Uffink
Milan, Italy — Giovanni Valente
Salzburg, Austria — Charlotte Werndl
Bristol, UK — Lena Zuchowski

Introduction

Tatiana Afanassjewa (1876–1964) was a Russian-Dutch mathematician and physicist, who made important contributions to the foundations and philosophy of physics. She was also a prominent voice in the didactics of mathematics and an active participant in some of the most influential intellectual debates of the earliest twentieth century. However, her legacy has received little attention from philosophers and historians of science: all too often she is remembered only as the lesser known co-author of the publications she wrote together with her husband Paul Ehrenfest (1880–1933) on statistical physics. While these influential collaborative works are part of her legacy (and will be discussed in this book), Afanassjewa's independent contributions, in particular to the foundations of thermodynamics and the didactics of mathematics, offer many visionary insights and deserve more exploration than has so far been accorded to them.

This book aims to provide an in-depth and comprehensive exploration of Tatiana Afanassjewa's legacy. We hope that it will (i) highlight Afanassjewa's independent work, thereby raising her profile in the philosophy of physics community and making sure that her achievements are not unjustly overshadowed by those of her husband, and (ii) analyse selected aspects of her works and demonstrate how they continue to yield insights into the foundations of physics and mathematics.

The book is an edited volume of original contributions from a diverse set of authors. A number of the papers collected in this book are based on contributions to the workshop *Tatiana Afanassjewa and her legacy: New perspectives on irreversibility*, which took place on June 17–18, 2017, at the University of Salzburg. However, we have also elicited additional contributions on aspects of Tatiana Afanassjewa's work and life that were not represented at the workshop. Furthermore, in order to make her work more accessible to physicists, philosophers and mathematicians, the volume will contain translations of key passages from publications that are currently only available in German and Dutch. The authors contributing to this book are all well-regarded experts in their relevant fields and we have been fortunate in attracting such a high-calibre field of contributors.

The book is divided into three parts: Part I (Chaps. 1–3) discusses Tatiana Afanassjewa's biography and independent works; Part II (Chaps. 4–6) presents select aspects of her collaborative work with Paul Ehrenfest (in this introduction, we follow the usual naming convention for the authors of these joint works by referring to Paul and Tatiana as 'the Ehrenfests'); Part III (Chap. 7–8) contains translations of Tatiana Afanassjewa's work on the foundations of thermodynamics, which is currently only available in German, and her publications on the didactics of mathematics, which are currently only available in Dutch. In the following, we will briefly introduce each chapter.

Part I: Tatiana Afanassjewa's Life and Forgotten Legacy

In Chap. 1, Margriet van der Heijden provides a biographical sketch of Tatiana Afanassjewa's life. Afanassjewa studied mathematics and physics both at the *Bestuzhev* courses for women and later at the 'regular' university in St Petersburg. In 1902 she went to Göttingen to study under Felix Klein, where she met Paul Ehrenfest, who shared her ideals, and who she married in Vienna in 1904. When the couple moved to St Petersburg, she became a prominent participant in debates on mathematical education there. Her life changed again when she accompanied Ehrenfest to Leiden, The Netherlands, in 1912, where he became a full professor of theoretical physics, as the successor of H. A. Lorentz, while laws and unwritten rules prevented her from obtaining an academic position. It demonstrates Afanassjewa's strength that she soon initiated a fierce debate on mathematical education—it led to the foundation of the academic journal *Euclides* for Dutch mathematics teachers—while also designing a house that would provide a welcoming household where Einstein, Bohr and at least a dozen of other Nobel laureates and many more academics and students participated in lively debates.

In Chap. 2, Marianna Antonutti Marfori explores Afanassjewa's work on the pedagogy of mathematics, in particular geometry, and discusses Afanassjewa's views on the teaching of geometry in the context of the early twentieth century debate on mathematical education. Afanassjewa holds that the educational value of geometry lies in its method and its quest for utmost clarity. By learning to process spatial images in their mind by representing them visually, filling in gaps, and identifying contradictions, the student can make the method of geometry their own, and go on to apply it to new problems, both inside and outside geometry. Both of the dominant approaches at the time, on Afanassjewa's view, fail to recognize this essential aspect of geometry. According to the first of these, geometry should be taught by laying out rigorous, discursive proofs in the style of Euclid. According to the second, geometry should be taught by developing insights arising from concrete examples. Since a rigorous, axiomatic presentation of the results of geometry does not show the thought process that brought it about, the student cannot understand or appreciate the importance of a logical presentation of geometry until they have already attained a certain mastery of the subject matter. On the other hand, the untrained student cannot generally be expected to make the correct generalizations from concrete examples. Afanassjewa argues that the correct approach is to develop the student's reasoning about spatial relations and their presentation, thereby also training the student's ability to reason logically. Once space has been studied systematically in this way, the student will be able to recognize the axioms of geometry as evident and appreciate the value of an axiomatic presentation of the subject. The chapter discusses the points of contact between Afanassjewa's views on the roles of logic and intuition in geometry and those of Poincaré, Klein, and Hilbert.

In Chap. 3, Jos Uffink and Giovanni Valente discuss crucial aspects of Afanassjewa's (1925, 1956) original work on the foundations of thermodynamics. First, they focus on her treatment of reversibility in thermodynamics and her introduction of 'quasiprocesses' in this treatment and show how closely this discussion relates to some current discussions in philosophy of science and show how her approach resolves a paradox put forward by Norton (2013, 2016) that allegedly plagues thermodynamically reversible processes. Another issue raised by Afanassjewa is whether, owing to the formal analogy between temperature and pressure as integrating divisors for heat and work, respectively, one could formulate the Second Law not just in terms of entropy, but also in terms of volume non-decreasing processes when no work is performed on a system. Yet, she pointed out that one can construct examples where the analogy breaks down, unless some extra axiom is added. Finally, Uffink and Valente take up her discussion of the alleged logical equivalence between Kelvin and Clausius formulations of the Second Law, which Afanassjewa questioned in light of the possibility of absolute negative temperature, 30 years before Ramsey (1957) made that possibility more widely known to the physical community.

Part II: The Ehrenfests' Work on the Foundations of Statistical Mechanics

In Chap. 4, Roman Frigg and Charlotte Werndl analyse the Ehrenfests' argument for the conclusion that the phase averages generated by Gibbsian statistical mechanics and Boltzmannian equilibrium values should coincide. The relation between the Boltzmannian and Gibbsian formulation of statistical mechanics is still a major conceptual theme in the foundations of statistical mechanics: therefore, the argument is still highly relevant today. The chapter fills in some important details the original argument skipped over and points out that the its scope is limited to dilute gases. This is not a shortcoming of their argument but an inherent limitation of the claim: it is not generally the case that Boltzmannian equilibrium values and Gibbsian phase averages agree. They then discuss the example of the six-vertex model and show that in that model the two values come apart and go on to offer a general theorem providing conditions for the equivalence of Boltzmannian equilibrium values and Gibbsian phase averages.

In Chap. 5, Patricia Palacios analyses the 'ergodic hypothesis' which the Ehrenfests prominently introduced and highlighted in their celebrated joint Encyclopedia article of 1911 as a crucial assumption of Boltzmann's approach to statistical mechanics. This article has been strongly criticized by historians of science as not providing an historically accurate account of Boltzmann's approach. However, Palacios also evaluates the role that the ergodic hypothesis of the Ehrenfests came to play in the subsequent development of ergodic theory in the course of the twentieth century and argues that the major constructive role of the Ehrenfest's discussion of the ergodic hypothesis in these developments stems precisely from those aspects about their formulation of the hypothesis that historians have regarded as historically inaccurate.

In Chap. 6, Joshua Lucasz and Lena Zuchowski highlight and discuss the Ehrenfests' use of toy models to explore irreversibility in statistical mechanics. In particular, the chapter explores their urn and *P–Q* models and emphasizes that while the former was primarily used to provide a simple counter-example to Zermelo's objection to Boltzmann's statistical mechanical under-pinning of the Second Law of Thermodynamics, the latter was intended to highlight the role and importance of the *Stosszahlansatz* as a cause of the tendency of systems to exhibit entropy increase. They also explain the sense in which these models are toy models and why agents can use them, as the Ehrenfests did, to carry out this important conceptual work, despite the fact that they do not represent any real system.

Part III: Translations from German and Dutch

Chapter 7 presents the translation by Marina Baldissera Pacchetti of one paper and four chapters of Tatiana Afanassjewa's book on the foundations of thermodynamics. The paper, published in 1925 in the journal *Zeitschrift für Physik,* is titled 'On the Axiomatization of the Second Law of Thermodynamics'. In this paper, Afanassjewa considers the axiomatic derivation of the Second Law of Thermodynamics by Carathéodory (1909) and argues that this derivation requires at least two more logically independent axioms. After 1925, she wrote many more papers on the foundations of thermodynamics and summarized her views in a book manuscript in the early 1940s. This book, entitled *Die Grundlagen der Thermodynamik* (The Foundations of Thermodynamics) was finally published in Leiden in 1956. This volume will provide a first translation of selected parts of this book in English. This translation includes the foreword, in which Afanassjewa clarifies her approach; the Chap. 1, in which she clarifies her use and understanding of fundamental terminology; Chap. 6, in which she discusses the distinction between processes and quasiprocesses and related issues—such as reversibility and entropy; Chap. 8, on the Clausius-Thomson principle and irreversibility; and, finally, the third appendix, in which she comments on the Bolzmannian H-theorem.

In Chap. 8, Pauline van Wriest translates Tatiana Afanassjewa's famous manifesto, *What can and should geometry education offer a non-mathematician?* (1924)

from Dutch, a manifesto which led to an intense dispute with E. J. Dijksterhuis on mathematical education which in turn led to the foundation of a new Dutch-language journal *Euclides*, devoted to the teaching of mathematics.

Jos Uffink
Giovanni Valente
Charlotte Werndl
Lena Zuchowski

Contents

Part I
Tatiana Afanassjewa's Life and Forgotten Legacy

Chapter 1
Tatiana Ehrenfest-Afanassjewa: No Talent for Subservience

Margriet van der Heijden

One tall house stands out at the end of Leiden's quiet and inconspicuous Witte Rozenstraat, just outside the city center. Its white-plastered walls, its size, and neoclassical design contrast with the modest appearances of the neighboring brick houses. A curious passer-by might notice the two plaques in the almost windowless wall on the street side. One is dedicated to the Austrian–Dutch physicist Paul Ehrenfest. "Here lived and worked professor Paul Ehrenfest," it says, simply. The white stone plaque was a gift from the *Christiaan Huygens Dispuut*, the debating society for students in mathematics and natural sciences in Leiden with a long and impressive history.[1] A second, similar plaque commemorates "His wife, Tatiana Afanassjewa who, ahead of her times, opened up this house for people and ideas."[2]

A more inquisitive passer-by will, after some further research, observe two more things. First, the plaque dedicated to Afanassjewa was placed there several years after the one for Ehrenfest, as if all other people living in the house, including Afanassjewa, as well as the house itself, were initially only considered to be part of the backdrop against which Ehrenfest performed his outstanding work. One could argue that this is how things are done: we remember and commemorate those who perform, not those who assist them in their performance. This would be a valid point, were it not for the fact that Ehrenfest is not remembered primarily for his research achievements in theoretical physics, though they are important, but rather for the role he played as a "knowledge broker" and "catalyst" within that field.

[1] The Dutch "Dispuut Gezelschap Christiaan Huygens" was established at the end of the nineteenth century and still exists today.

[2] Tatiana Afanassjewa herself always used this German transcription of her name when not in Russia, both privately and when publishing essays and research articles. It will be used in the current article as well.

M. van der Heijden (✉)
Amsterdam University College, Amsterdam, Netherlands
e-mail: m.w.vanderheijden@auc.nl; margrietheijden@gmail.com

J. Uffink et al. (eds.), *The Legacy of Tatjana Afanassjewa*, Women in the History of Philosophy and Sciences 7, https://doi.org/10.1007/978-3-030-47971-8_1

Critical, excellent in spotting high-profile and groundbreaking work of colleagues and, as his brother once wrote, "with a flair for recognizing outstanding personalities with whom you then quickly get in touch,"[3] Ehrenfest was for theoretical physics what a charismatic gallery owner can be for the arts.[4] Colleagues like Einstein, Bohr, Sommerfeld, and many others greatly appreciated his ability to evaluate their work, to recognize its weak and strong points, and to link it to old and new trends in physics in crystal clear language.[5] They valued Ehrenfest's large network, his helpfulness in bringing people from different places and backgrounds together, and his absolute and conscientious dedication to physics. They loved to visit the large house at the end of Witte Rozenstraat, which served as a meeting place and, in a sense, a gallery for physics.

What role did Afanassjewa play in all that? A second observation a passer-by might make is that, before her name is even mentioned, she is defined in relation to Ehrenfest: she was "his wife."[6] Does that imply that her role, in "opening up the house," was a traditional one? Was she the professor's wife who enjoyed being a hostess for the many—mostly male—scientists that came to the house with a mind full of ideas and with high hopes of sharing and discussing those ideas with other visitors? In the context of Dutch society, with its conservative stance on gender roles, it would be tempting to answer these questions affirmatively, especially since no further details on Afanassjewa's background, training, or possible public roles are given.

Yet, in spite of the good intentions of the *Christiaan Huygens Dispuut*, such an interpretation does not do justice to Tatiana Ehrenfest-Afanassjewa, who did indeed receive plenty of guests in Leiden and elsewhere, but who had no talent for subservience. Afanassjewa was an independent and successful Russian mathematician and physicist in her own right, with the extraordinary courage to trace out her own path under circumstances and in times and places that hardly allowed women to develop and use their talents. She not only opened up her home to outsiders, as the plaque commemorates, she herself designed this magnificent house which, to this day, shows a number of interesting "Russian" details.[7]

---

[3]Hugo Ehrenfest to Paul Ehrenfest, 9 April 1924: Ehrenfest Archive, Museum Boerhaave Leiden (EA-MBL) 1.1.2.

[4]Ehrenfest himself inspired this notion, since he once compared Einstein to Holbein and Bohr to Rembrandt during a conversation with Robert Oppenheimer: Undated note from R. Oppenheimer to M. J. Klein, EA-MBL 12.1.

[5]Klein (1970).

[6]Marriage certificate, 21 December 1904: EA-MBL 2.2.

[7]Examples are thick walls that keep the house warm in winter and cool in summer; a heating system with horizontal pipes rather than radiators; double-glazed windows with large spaces between the inner and outer layers of glass; "lazy" stairs that are more likely to be found in Saint Petersburg than in Leiden. Sketches and building plans: Ehrenfest Family Archive (EFA).

1.1 The Alarming Rectilinearity of Her World line

Who was Tatiana Afanassjewa? In a letter to her and Ehrenfest, written shortly after a stay in Leiden, their friend Albert Einstein wrote jokingly: "T will also join us [in appreciating Bach] despite the alarming rectilinearity of her intellectual worldline (an exception to the laws of motion?)."[8] Einstein is referring here to Afanassjewa's lack of willingness, at least until then, to attach a higher value to Bach's chorales than to those by Russian composers. In the same stroke, he also characterizes her entire intellectual development as rectilinear.

Grudging admiration seems to resound in the little joke. Afanassjewa was totally different from the women that surrounded Einstein at the time. Einstein had just spent years finishing the covariant equations of relativity, after years of work,[9] and was in the midst of formalizing his divorce from Mileva Mariç. In addition, his future family-in-law was putting pressure on him to marry Elsa Einstein, who had been taking care of him and had already been waiting for him for quite some years.[10] They unnerved him, "these women" who "always wait for someone to come along who will use them as he sees fit,"[11] as he once wrote in these gloomy months. Afanassjewa offered a striking contrast: she had an independent streak, an analytical mind, she strongly expressed her ideas about education, and she freely participated in the many discussions about physics that took place in the large study at Witte Rozenstraat. Partly inspired by Leo Tolstoy whose portrait had a prominent place in the study,[12] she was also a vegetarian and abstained from alcohol and smoking, just like her husband.

Tolstoy's books and ideas had been part of her upbringing. Afanassjewa was raised by her aunt and uncle: the respectable and childless Sonya Maslova and her husband Pyotr Afanassjew, who worked as a chief engineer for the tsar's railways. Her mother, Yekaterina Ivanova, had taken little Tatiana from Kiev to Petersburg when she was only two years old, after her husband, the engineer Alexey Afanassjew, had suffered a major mental breakdown and had been committed to a mental asylum. In this city of tsars, ice, and white nights, a city oriented toward the Western world, her aunt and uncle had treated Afanassjewa as their own daughter and had given her an excellent education.[13]

A good education was one of the things Afanassjewa shared with Einstein's soon to be ex-wife wife Mariç, just as Ehrenfest shared quite a few traits with Einstein himself. Ehrenfest and Einstein were almost the same age, had both been raised in a secular, Jewish, middle-class family, both loved to play music, and both strongly

---

[8]Einstein to Ehrenfest, 18 October 1916: *The Collected Papers of Albert Einstein* (CPAE) 8a, Doc. 268; Also cited in Klein (n.5) 304.

[9]The first detailed discussions of this work can be found in letters to Ehrenfest: CPAE 8a, Docs. 182, 185.

[10]See, e.g., CPAE 8a, Introduction.

[11]Einstein to Besso, 21 July 1916: CPEA 8a, Doc. 238.

[12]Senger and Ooms (2007).

[13]Biographical notes by son-in-law Henk van Bommel, undated: EFA.

disliked the German educational system. Afanassjewa and Mariç were raised in families that adhered to the Orthodox Church—in Russia and Serbia, respectively––both were a couple of years older than their husbands, they had completed the gymnasium (a kind of grammar school) before studying physics and math, and both went abroad to study at university.[14]

Yet, there was a crucial dissimilarity between the two women as well. The rebellious and fierce Mariç had been crushed by the immense talents and ambitions of her Albert. In their household with two small sons, amidst the laundry and the cooking, Mariç had become a somber shadow of her former self.[15] By contrast, Afanassjewa and her charismatic, insecure, and restless Paul had managed to organize their household, eventually including four children, in such a way that Afanassjewa could continue to study and work—though perhaps not as much and as freely as she would have liked. It made an impression on Einstein, as he wrote after another stay in Leiden, three years later, in 1919: "Not in any other house did I experience such a joyful family life; it stems from two independent people who are not bound together by compromises!"[16]

Einstein had been equally observant when he had used the epithet "rectilinear" to characterize Afanassjewa's "intellectual worldline," as some persistent trends can be observed in her intellectual world. Guiding lines in her life were her rock-solid passion for mathematics and physics, particularly thermodynamics, the value she attached to independent, logical, and critical thinking, as well as her clear and strongly voiced ideas about education, especially about teaching geometry.[17] Another constant throughout her life was her attachment to things Russian: its hills and forests,[18] its music, language, and literature, as well as some of its traditions and many of its scientists.

1.2 Higher Women Courses for the Weaker Sex

What was Russia like while Afanassjewa grew up there? How did she end up in Leiden? Her link to this modest Dutch town was Ehrenfest, whom she met in Göttingen in 1902, ten years before he became successor to Hendrik Lorentz[19] at the University of Leiden. After finishing her studies in Russia, Afanassjewa had traveled to Göttingen, the German "Mecca of mathematics," with her aunt Sonya, hoping to deepen her knowledge of physics and mathematics. She was not the first Russian woman to do so. The two most important mathematicians in town, Felix Klein and

[14]Mariç went to Zürich immediately after high school. See, e.g., Popović (2003).

[15]See, e.g., CPAE 9, Introduction.

[16]Einstein to Ehrenfest, 9 November 1919: CPAE 9, Doc. 155.

[17]Ehrenfest-Afanassjewa (1960). Personal details: from the preface to this collection of essays by Dutch mathematician Bruno Ernst [pseudonym of J.A.F. [Hans] de Rijk].

[18]Personal communication T. van Bommel.

[19]Hendrik Antoon Lorentz (1853–1928). For a concise biography, see Kox (2018).

David Hilbert, had no objection to women in their field and in the years before Afanassjewa's arrival two Russian women, Lyubov Zapolskaya[20] and Nadezhda Gernet,[21] had finished their doctorates with Hilbert.[22]

Yet, Afanassjewa hardly received a warm welcome. German students in several cities had raised objections against the number of foreign students, and especially against *female* foreign students, for the most part Russian, who were taking up space and time in the already overcrowded lecture rooms. In response, the University of Leipzig and the medical faculty in Göttingen had decided not to admit women any longer and, although Hilbert and Klein did not follow this example, it is not surprising that the mathematics and physics students strictly adhered to the old policy of excluding women from their weekly colloquia.[23] "Women are invited during festivities only," was the reply Ehrenfest supposedly received, when he inquired why the new Russian student—Afanassjewa—had not been invited. Eventually, it was Ehrenfest who successfully challenged the unwritten rules. It marked the start of a long-term relationship between him and Afanassjewa.[24]

Afanassjewa was happy to participate, discuss, and learn, but: "There was a large difference between what the professors in Petersburg had taught us, and what was discussed in Göttingen (Klein, Hilbert, Minkowski)," she said, much later in life.[25] For women, nothing in the educational system was to be taken for granted and this was true as well for Afanassjewa, even though she had grown up in St Petersburg, where a feminist elite had advocated for higher education for women well before women in most European countries began to do so.[26] In St Petersburg, the first gymnasia for girls opened as early as the middle of the nineteenth century, and around 1860, many well-to-do citizens opened up their salons to women for free lectures on Sundays, while women also began to attend lectures at the university. After Tsar Alexander II had prohibited all these activities in 1862, an impressive number of Russian women went abroad to attend the universities of Zürich and Paris, which had just opened

[20]Lyubov Zapolskaya (1871–1943). Like Afanassjewa, she studied at the Pedagogical Institute, the Bestuzhev Institute, and then obtained her Ph.D. in Göttingen, with Hilbert (1901). In Russia, in Saratov, she then taught mathematics, among other subjects, and headed the department of Higher Mathematics and Mechanics at the Pedagogical Institute of Yaroslav. *she-win.ru/nauka/588-lubov-zapolskaya.*

[21]Nadezhda Gernet (1877–1943) first studied at the Bestuzhev Institute, and then obtained a Ph.D. with Hilbert in Göttingen (1902). She became a teacher at Bestuzhev, and went on to teach at the university, once its courses had merged with those at the Bestuzhev Institute. In 1930, she became a professor at the Polytechnic Institute in Petersburg: Editors A.N. Kolmogorov and A.P. Yushkevich, *Mathematics of the 19th century* (Basel 1998).

[22]Klein received support for his policy from the Prussian minister for education. E.g., Thiele (2011).

[23]Bonner (1995).

[24]Afanassjewa to M. J. Klein, undated: EA-MBL 12.1.

[25]Ehrenfest-Afanassjewa (n.17). Strikes and unrest in 1899 may have had a negative effect on the courses.

[26]The cities were different from the rural areas in Russia, where the majority of the population was illiterate.

their doors to women students. In fact, of the 203 women who studied in Zürich between 1864 and 1872, 148 came from Russia.[27]

Afanassjewa's generation was the first one to profit from their efforts. Fearing that, after their return from abroad, women scientists and lawyers would oppose his policies and undermine his authority even more, Alexander II and his government relented and allowed for the establishment of Higher Women Courses, in other words, higher education for women. The Higher Women Courses in St Petersburg opened their doors in 1878, just before Afanassjewa first arrived in the city. They were soon referred to as the Bestuzhev courses, after their first dean, the historian K. N. Bestuzhev-Ryumin.[28] A private organization took care of funding for the institute, raising money through book sales, concerts, and by collecting gifts,[29] and not long afterward a large new building appeared on the Vasilyevsky Island, not far from the university.[30]

Large efforts were made to guarantee high-level lectures. Professors like Dmitri Mendeleyev and Alexander Borodin taught courses at the Bestuzhev for a small fee. Other scientists offered moral support: "If a woman receives an adequate education and training she can pursue culture in science, art and public life just as well as a man,", the brilliant professor and surgeon Nikolay Pirogov wrote to a baroness friend of his, in the year when Afanassjewa was born.[31] This intellectual climate, as well as having been raised in an academic environment (her uncle would soon become a professor of mathematics at the Polytechnic Institute in Petersburg,[32] while her aunt was in favor of modern education[33]) seemed to almost predestine Afanassjewa to attend the Bestuzhev courses. Yet, her uncle sent her instead to a pedagogical institute that trained teachers up to the lower levels of the gymnasium.[34] Was he afraid perhaps––with his brother in mind—that Afanassjewa's nerves would suffer from intensive studying? Or was he deterred by the reputation of the Bestuzhev Institute, which was

[27] Koblitz (2013).

[28] www.prlib.ru/en/history/619592. Women could study History and Philology or Physics and Mathematics. After 1906, a third possibility was Law.

[29] Stites (1978).

[30] Currently the faculty of Earth Sciences of the University in Petersburg.

[31] Hans (1963). These words were not empty: As early as 1864 9,000 girls were enrolled in 29 girls' schools of the first order (later called gymnasium) and 91 of the second order (later called progymnasium) and in 1869 another 32 girls' schools had been established.

[32] Van Bommel (n.13).

[33] Sonya had sent Afanassjewa to a new private gymnasium, in a large building at the Ulitsa Kabinetskaya, slightly south of the Fontanka. The director, Maria Nikolyevna Stoyunina, applied the innovative pedagogical principles her husband, Vladimir Stoyunin, had described in lectures and books: teachers tried to foster individual talents, pupils were allowed to jump, run, and talk for 15 min between classes to refresh their minds, and they had gymnastics classes every day. Latin and Greek were not taught; the curriculum was a watered-down version of what boys were taught, according to what Afanassjewa said later [n. 17], "but the pedagogical methods were such as I would like to see them everywhere."

[34] Called Pedagogical Courses of the Girls' Gymnasium and, from 1903 on, Women's Pedagogical Institute.

increasingly seen as rebellious, and even as a "hotbed of anarchism," according to the Okhrana, the tsar's secret service?[35]

Afanassjewa dutifully, but regretfully, finished her studies at the pedagogical institute in 1897 and then accompanied her aunt on a trip that finished in Vienna at Christmas. Little did they know that Afanassjewa's future husband—a somber adolescent who had recently lost both his parents—was a student at the Kaiser Franz Joseph Gymnasium there.[36] Their stay in the city was marred by the message they received that Afanassjewa's uncle Pyotr was seriously ill. He died soon afterward and a few weeks later, Afanassjewa enrolled in the Bestuzhev, halfway through the first year, with financial support from her aunt.[37]

The next three years she spent enjoying the lectures and working hard.[38] The lack of enthusiasm of her physics teacher, Orest Khvolson, for her work in her favorite field disappointed her.[39] On the other hand, she was encouraged by her favorite math teacher, Vera Iosifovna Shiff.[40] "I was very pleased to read in the newspaper that the Tsar has decreed that the university of Helsingfors [Helsinki] has to admit women on the same conditions as men," Shiff wrote to Afanassjewa in the summer of 1902. "Let us see whether women will be admitted to Russian universities as well. I cannot think of any logical reason why women and men would not attend university together."

"Here in Germany for example [Shiff spent her vacations in Germany] where society, and students especially, are not in the least sympathetic to the idea of higher education for women, women are nevertheless admitted to universities without causing any problems," she added, and her following remark may have bolstered Afanassjewa's self-confidence. "Yet people say that our girls' gymnasia do not meet the requirements of the university, while Germany only has two girls' gymnasia, one in Karlsruhe and one in Stuttgart, and German girls usually visit the Höhere Mädchenschule, which has a level **way below** that of our Mariinskiye gymnasia."[41]

---

[35]Richard Stites (n.30); Morrissey (1998).

[36]Johanna Ehrenfest-Jellinek died 3 May 1892; Sigmund Ehrenfest died 10 November 1896.

[37]Van Bommel (n.13).

[38]In 1899, large-scale strikes and student protests took place, and they can be viewed as a prelude to the revolution of 1905, although in 1899 the students were still more preoccupied with academia and their own position than with society at large. Afanassjewa did not participate in these protests: she was among the women who believed in more gradual change and who were afraid that the Women's courses would be closed in retaliation to protests. See Morrisey (n.34).

[39]Orest Chvolson (1852–1934) was a physics professor in Petersburg and was well known for his textbooks, which were translated into many languages.

[40]Shiff herself had completed her studies at the Women's Courses in 1882. She had later obtained a Ph.D. in Göttingen and had become a teacher at the Women's courses. She taught geometry and calculus, amongst other subjects, and published academic books and articles in Russian academic journals; she was one of the first Russian women who dedicated her life to mathematics: Mary (2015).

[41]Vera Shiff to Afanassjewa, 26 June/9 July 1901, EFA [translated from Russian to Dutch by Hans Driessen].

1.3 Afanassjewa's Analytical Mind and Ehrenfest's Physics Intuition

Did Afanassjewa's talents and training match her ambitions, as she went from Petersburg to Göttingen, from Göttingen to Vienna, and from Vienna back to Russia? In Göttingen, she felt intimidated, as she later confessed: "The people there had completed their formal studies at other places in the world. Most of them were younger than I was and knew more."[42] In Vienna, where she married Ehrenfest in December 1904, and where she attended Boltzmann's lectures as an auditor, she felt completely out of place.[43] And when she and Ehrenfest returned to Göttingen, in 1906, living on their inheritances, Afanassjewa had little hope of starting on an academic career. By then she was almost 30 years old, her diploma from the Women's Courses was looked down upon by many academics, and—in contrast to Gernet, Zapolskaya, and Shiff—she had not managed to obtain a doctorate in Germany. To complicate matters, she had a daughter, Tanichka, who was born in Vienna at the end of 1905.

Still, perhaps indirectly, she must have made an impression on Klein. The dignified Prussian mathematician had, among many other things, set himself the task of compiling all recent insights from the field of mathematics and physics in an *Encyklopädie der mathematischen Wissenschaften.* **Statistical mechanics, based on the monumental work of the famous Boltzmann, was one of the fields to be included, and** Klein asked Ehrenfest, a former student of the recently deceased Boltzmann,[44] whether he would be willing to write this particular contribution to his encyclopedia—"possibly with your wife." [...] "In any case I would like to ask you and your wife to visit me at home, perhaps tomorrow (Sunday night) at 6 o'clock, to discuss matters."[45]

Klein's request did not come out of the blue. Just days earlier, at the weekly meeting of the *Mathematische Gesellschaft*, Ehrenfest had presented the urn model[46] that he and Afanassjewa had developed to illustrate and discuss two major objections against Boltzmann's famous H-theorem.[47] During her life Afanassjewa would often be ahead of her time—with her novel ideas on didactics, for example—or lag behind, like when she arrived in Göttingen just after Gernet. Yet, the timing of this presentation was perfect. It confirmed Klein's impression that Ehrenfest had a sound physics intuition and Afanassjewa a sharp analytical mind, and it convinced Klein that they were both well acquainted with Boltzmann's work. By assigning them this

[42]Ehrenfest-Afanassjewa (n.17).

[43]Klein (n.5); Auditor-papers university Vienna: EA-MBL 2.4.

[44]Ludwig Boltzmann (1844–1906), who had been suffering from depression for many years, committed suicide during a vacation in Duino, Italy, in September 1906.

[45]Klein to Ehrenfest, 16 November 1906: EA-MBL 1.2.2.

[46]Later also referred to as the "flea model."

[47]Paul had presented their "Urnmodel" at the *Mathematische Gesellschaft* at the beginning of November 1906. It later appeared as P. and T. Ehrenfest, "Über zwei bekannte Einwande gegen das Boltzmannsche H-Theorem," *Physikalische Zeitschrift* 8 (1907) p. 311.

task, Klein thus gave two unemployed physicists—one a doctor of science, the other with a somewhat mysterious status—the chance to participate in a prestigious project and have their work published side by side with the writings of famous scientists like Arnold Sommerfeld, David Hilbert, and Carl Runge.[48]

1.4 Undervalued: Kruzhoks and Odd Jobs

In the following years, similar small tokens of support at crucial moments helped Afanassjewa advance in mathematics and physics in St. Petersburg, where they had moved to in 1907, after failing to secure a position in Göttingen. Much has already been written about Ehrenfest, who befriended Abram Ioffe[49] and—at the instigation of Afanassjewa's aunt—began to organize biweekly *kruzhoks*[50] where recent papers in the field of physics were discussed. He quickly became one of the editors of the *Journal of the Russian Physicochemical Society*,[51] and, with his solid training in Vienna, Göttingen and, for a brief three months, in Leiden,[52] he was soon viewed as an authority and a superb guide in the novel field of theoretical physics in Petersburg.[53]

Less is known about how Afanassjewa held down a variety of odd jobs to make ends meet while organizing her own *kruzhok* on probability theory and complementing her earlier studies with a degree in physics and mathematics from the University, which had just opened its doors to women. Actually, just a couple of years later, the diplomas from the women courses and universities would be considered equivalent and the two institutions would merge, so her timing was—once again—far from fortunate.[54] In the first year, after their arrival in Petersburg, she worked as a mathematics teacher at a private *gymnasium* [grammar school] and—for a few hours each week—as an assistant teacher at the Bestuzhev Courses. In the fall of 1908, after moving to Apothecary Island, a bit further away from the center, she secured for herself a small job at the Pedagogical Museum of the Military Academy. This institution offered lectures to recruits, and to others who were interested, on a variety of topics and a lecture that Afanassjewa herself had proposed—on Hilbert and the foundations of geometry—was well received. The director of the Museum,

[48]P. and T. Ehrenfest, "Begriffliche Grundlagen der statistischen Auffassung in der Mechanik," *Encyklopädie der mathematischen Wissenschaften* IV (Leipzig 1911). In English: *The conceptual foundations of the statistical approach in mechanics* (Ithaca 1959).

[49]Abram Ioffe (1880–1960) obtained his Ph.D. with Wilhelm Röntgen in Germany and then moved back to Petersburg where he would later, in Soviet times, become the director of the Leningrad Physico-Technical Institute (LPTI).

[50]Diary and notebooks: EA-MBL 1.2.6; Klein (n.5). Kruzhok means "small circle": the papers were discussed within a small circle of physicists and mathematicians.

[51]Frenkel (1977) Chapter 3; Tsipenyuk (1973).

[52]Ehrenfest spent some time in Leiden in the spring of 1903: notebooks 1903: EA-MBL 1.2.6; Klein, *Ehrenfest* (n.5).

[53]Frenkel (n.51).

[54]Koblitz (n.27).

General and mathematician Zakhar Maksheyev, who had "accepted her offer to give a lecture with gratitude,"[55] was so impressed that he asked Afanassjewa to become a member of his working group to discuss, among other things, reforms in mathematics education.[56]

Such reforms were a hot topic in those years, both in Russia and abroad.[57] There were two opposing camps, as Afanassjewa later wrote in an essay about the didactics of geometry.[58] The "logicians," on the one hand, were in favor of teaching rigorous proofs: students should learn to derive all theorems from a set of axioms, just like Euclid himself. The "pragmatists," on the other hand, preferred to illustrate the validity and relevance of the various theorems by using concrete examples, preferably from everyday life. Since each camp claimed that its own method was the most successful one, using examples from its own classrooms only, Afanassjewa in 1909 proposed to directly ask first-year students how they looked back on *their own* mathematics education in high school. She emphasized that students in all types of higher education should participate: those at the conservatory, the academy of arts, as well as the university. In the summer of 1910, at the request of Maksheyev, she drew up a questionnaire with 54 questions about mathematics in general, algebra, geometry, and trigonometry, which was to be sent to 10,000 students.[59] She was even working on it in a dacha in Kanuka, only a few days after having given birth to her second daughter, Galinka, and while Ehrenfest was away, discussing his future with his friend Herglotz[60] in Villa Hedonia, a spa in Bad Kissingen.[61]

In the following year Afanassjewa also plunged deeper into physics. The proceedings of the *First All-Russian Congress for Mathematics Teachers* at the end of 1911 in Petersburg, where she lectured on irrational numbers and participated in the discussions as well as in the publications in the *Journal of the Russian Physicochemical Society/JETP*, show that her voice began to be heard.[62] At the same time, Ehrenfest became more and more unhappy in Russia. Since mixed marriages were not allowed in Austria, he and Afanassjewa had given up their religious affiliations in 1904, and being a "non-religious man" made it almost impossible for him to obtain an academic position. That he was of Jewish origin did not help, neither did his critical attitude

[55]Maksheyev to Afanassjewa, 1 April 1908: EFA [translated from Russian to Dutch by Hans Driessen].

[56]Ehrenfest-Afanassjewa (n.17).

[57]Weigand et al. (2017).

[58]Ehrenfest-Afanassjewa (n.17) Chapter III.

[59]Undated handwritten manuscript of the questionnaire: EFA [translated from Russian to Dutch by Hans Driessen].

[60]Gustav Herglotz (1881–1953).

[61]Correspondence Ehrenfest and Afanassjewa, July and August 1910, EA-MBL 1.1.2; Diary, 23 July 1910 [translated from Russian to Dutch by Hans Driessen]; Pim Huijnen, *Die Grenze des Pathologischen, het leven van fysicus Paul Ehrenfest, 1904–1912* (Master's Thesis University of Groningen, 2003).

[62]Proceedings of the *First All-Russian Congress for Mathematics Teachers*, 27 December 1911–3 January 1912 [relevant parts translated from Russian to Dutch by Hans Driessen]; Ehrenfest-Afanassjewa (1911a, 1911b, 1912) [translation ibidem.].

toward authorities in academia, even though he had established himself informally as a key player in the field of theoretical physics in Petersburg. "He is the man of the hour," Ioffe wrote to his wife after Ehrenfest's contribution to the *Twelfth All-Russian Congress of Natural Scientists and Medical Doctors* in Moscow between Christmas and New Year 1911/1912.[63]

1.5 A Vibrant Household in Traditional Leiden

Nine months later, Afanassjewa took her interest in thermodynamics, statistical mechanics, probability theory, and mathematics education with her to The Netherlands, where Ehrenfest, to the surprise of many, none the least himself, had been appointed as Lorentz's successor.[64] Little is known about the reasons why Lorentz, who had initially hoped to appoint Einstein to Leiden, asked Ehrenfest, but it is clear that he had been quite impressed by Ehrenfest's and Afanassjewa's long paper on statistical mechanics in the Encyklopädie.[65] It is also clear that Ehrenfest was overjoyed: his lack of prospects in Russia, as well as the grim future he foresaw for Russia itself, had increasingly been depressing him.

In contrast, for Afanassjewa, it was a painful farewell to the city she loved, where she had built herself a life with friends and *kruzhoks* and where her daughters were surrounded by relatives and where she had hoped, for example, to finalize the results from the questionnaire survey in which she was so invested.[66] In the first days of 1914,[67] she already traveled back to Moscow to participate in the second All-Russian Congress for Mathematics Teachers, much to the chagrin of her husband. "Try not to see your exile to Western-Europe in a dark light," he wrote to her. "Think about these three things: (1) that an infinite amount of urgent work needs to be done in Russia, but why? Because no one lets anyone do this work (2) You, and especially you, are not able to solve this problem (3) If only, if only you would be willing to proceed from the right angle, you could do many good things here [in Leiden] and in doing so develop yourself and exert lasting influence—not in the least by educating your children and through reading and writing."[68]

His concern was unnecessary: Afanassjewa returned and soon afterward the First World War, and later the revolution in Russia, would prevent her from visiting her home country for almost a decade. She was safe in an unfamiliar, flat country that

[63] Frenkel (n.49).

[64] Ehrenfest had visited Leiden in 1903 (n.53) but the two men had not met each other since, and Lorentz was under the impression that Ehrenfest was a professor in St. Petersburg now (Klein, n.5).

[65] Ehrenfest & Ehrenfest (n. 48).

[66] Afanassjewa discussed a.o. a very preliminary report about the survey during the second All-Russian Congress for Mathematics Teachers in Moscow, early in 1914, but the project stagnated during the turmoil of the First World War, as she herself stated Ehrenfest-Afanassjewa (n.17).

[67] In Russia it was still December 1913.

[68] Ehrenfest to Afanassjewa, 9 January 1914, 11 pm: EA-MBL 1.1.1.

was democratic and peaceful. From a different perspective, she was also trapped in this flat country, where women disappeared into the kitchen or the living room and, in any case, behind the front door, as soon as they were married. Most Dutch looked down upon women with paid jobs, pitying them for having been unable to find a supporting husband. The rather patronizing activities of the feminist elite in Leiden reflected this attitude: one of these activities was to provide needy women with an extra income by allowing them to do paid needlework—at home, unseen and anonymous.[69] That women obtained the right to vote, in 1919, did little to change this oppressive atmosphere. In 1924, a new law was even passed to formally exclude married women from civil service jobs and from teaching positions.[70]

Unsurprisingly, Dutch universities were predominantly male, though some female students had been admitted to the physics department in Leiden. Just before Afanassjewa and Ehrenfest arrived in this rather provincial town, one of them even defended her doctoral dissertation: Berta De Haas-Lorentz.[71] She was the eldest daughter of the eminent Lorentz, who in later years would become Afanassjewa's sparring partner in discussions about thermodynamics.[72] Between 1910 and 1920 Lorentz supervised three more women: Johanna Reudler, who obtained her degree in 1912, Eva Dina Bruins, who did so in 1918, and Hendrika van Leeuwen, who followed in 1919. However, only the latter would remain unmarried and would go on to pursue a career in academia.[73] That Lorentz accompanied his wife, Aletta Lorentz-Kaiser, to meetings about women's rights, also illustrates that he was unusual and more modern than many of his Dutch contemporaries.[74] It explains why he took an interest—though a modest one!—in Afanassjewa's capabilities.

A few days after their arrival in Leiden, Lorentz sent Ehrenfest a postcard: "I forgot to ask you yesterday whether Mrs. Ehrenfest, if she is not too busy with furnishing the house, would like to visit the colloquium tomorrow night. It would be a great pleasure to me if she would accept this invitation."[75] Soon afterward, he introduced Afanassjewa to Rommert Casimir, founder and director of the Nederlandsch Lyceum[76] in The Hague, who was strongly in favor of innovative pedagogical methods. Through

69 Steen (2011).

70 De Haan (1998).

71 De Haas-Lorentz (1913); Original: *Over de theorie van de Brown'schen beweging en daarmede verwante verschijnselen* (Ph.D. Dissertation, University of Leiden 1912).

72 A postcard from De Haas-Lorentz to Afanassjewa, 2 May 1956, also refers to "the many hours in which we discussed thermodynamics together and during which I learned so much from you": EA-MBL 2.2.9:482.

73 Hendrika van Leeuwen (1887–1974) coordinated the physics lab for students at the Technical University in Delft, carried out research into magnetism, and became the first female "lector" at this Technical University in 1947. See, e.g., Blaauboer et al. (2015).

74 Personal communication A.J. Kox.

75 Lorentz to Ehrenfest, 12 November 1912: EA-MBL: 1.2.2.

76 Rommert Casimir (1877–1957) was also extraordinary professor of pedagogy at Leiden University between 1918 and 1947. At the Nederlandsch Lyceum, the first school of its kind, where, after two preparatory years, students could choose to focus on science (HBS) or opt for the curriculum of a traditional gymnasium.

him, Afanassjewa established a small network of mathematics teachers that began to meet monthly at Witte Rozenstraat to discuss reforms in mathematics education.[77] At the suggestion of Casimir, she also adapted and translated from Russian to Dutch some of her essays on this topic.[78]

Through reading and writing and by attending the Wednesday evening colloquia, organized by her charismatic husband, as well as by participating in the many discussions with guests, Afanassjewa did indeed develop her intellectual interests further.[79] While the war raged through Europe and while Russia suffered in the years after the revolution, she also gave birth to two sons—Pawlik (1915) and Vassily (1918)—and played a crucial role in creating a home that was open to people and ideas. Following Einstein, who had befriended Ehrenfest in 1912, a growing number of other luminaries—Niels Bohr, Paul Dirac, Wolfgang Pauli, Werner Heisenberg, Erwin Schrödinger, Lise Meitner, Robert Oppenheimer, and many more—found their way to the vibrant household. Their presence created a stimulating atmosphere and enlivened the small and rather reserved academic community in Leiden, which was home to Nobel laureates Kamerlingh Onnes and Lorentz, to the well-respected physicists De Haas and Kuenen, to the renowned astronomer de Sitter, and to bright students like Burgers, Struik, Coster, Kramers, Uhlenbeck,, Goudsmit, Tinbergen, and Casimir.[80] Visitors and students flocked to the large study at Witte Rozenstraat and the signatures on the wall in the attic where guests used to sleep still bear witness to those joyous and stimulating years.[81]

1.6 The First Cracks in the Relationship

Happiness hardly ever lasts long, and in the 1920s, the first cracks appeared in Ehrenfest's and Afanassjewa's happy alliance. After the Russian Revolution of 1917, Afanassjewa's shares in the Russian Railways were no longer of any value and could no longer serve as collateral for the huge mortgage on the large house. Moreover,

[77]Ehrenfest-Afanassjewa (n.17).

[78]The first translated essay was "*Over de rol der axioma's en bewijzen in de meetkunde*" in 1915: Ehrenfest-Afanassjewa (n.17).

[79]She published one more article in Russia in those years: Ehrenfest-Afanassjewa (1914).

[80]Theoretical physicist Hendrik Kramers (1894–952) worked with Bohr and later became professor in Utrecht, Leiden, and Delft; Physicist Jan Burgers (1895–1981) became professor in Delft and Maryland; Mathematician Dirk Struik (1894–2000) became professor at MIT; Physicist and meteorologist Dirk Coster (1889–1950) became professor in Groningen (he helped Lise Meitner escape from Nazi Germany); Samuel Goudsmit (1902–1987) was, among other things, head of the ALSOS mission of the Manhattan Project and worked in Brookhaven National Laboratory. Earlier, in Leiden, he had discovered electron spin together with George Uhlenbeck (1900–1988), who became professor in Ann Arbor and at the Rockefeller University in New York; Physicist and economist Jan Tinbergen (1903–994), who turned to economics at Ehrenfest's instigation, was awarded a Nobel prize (1969); Theoretical physicist Hendrik Casimir (1909–2000) became director of the Philips Physics Laboratory (Natlab).

[81]Hollestelle (2011); Van der Heijden (2015).

the care for their youngest son Vassily, a little boy with Down syndrome who had been placed in a "modern" institution in Jena, Germany, was expensive.[82] More and more frequently, Ehrenfest complained that the financial burden forced him to take on extra work—like giving lectures and serving as an examiner elsewhere—which kept him from being creative. Increasingly, Ehrenfest also emphasized that he felt unworthy of occupying Lorentz's former chair, especially when young "Knaben" like Heisenberg and Dirac began to revolutionize quantum physics. In a long letter that he intended to read aloud to Afanassjewa, he vented his frustration by blaming her strong ties to former Russian colleagues and friends for his situation. He began calmly, by stating that it was "very understandable," that a big bunch of Russians, including "some asthmatic Russian professor" [...] "turn to us." He drew up a list ranging from her aunt to Friedmann[83] and then went on to complain: "what a terrible pressure it puts on me, someone incapable of handling all these things," questioning his need to "sacrifice [his] own development."[84]

Afanassjewa was not the type of person to reciprocate and vent her own frustration in long letters or emotional outbursts. Yet, the years after 1917 had not always been easy for her either. She had been deeply worried about Vassily (Vassik), her youngest, and she had been agonizing over the fate of her mother and friends in the young and isolated Soviet Union. Like Ehrenfest, she had fretted over their expenses, in particular during and directly after the war when food was still scarce and rationed. Under these circumstances, finding time for studying and writing had been difficult. The situation changed after Afanassjewa had taken Vassik to the institution in Jena, in the spring of 1922. Four months later, to her great relief, her mother finally managed to travel to the Netherlands.[85] "Baba Katya," as she was called in Leiden, soon began to take care of the other three children. It marked the beginning of a period during which Afanassjewa was finally able to study, write, and think again.

1.7 Novel Ideas, not Always Appreciated

Two years later, in 1924, Afanassjewa caused a stir with her essay *What can and should geometry teaching offer a non-mathematician?*[86] Neither siding with the "logicians," nor with the "pragmatists," she proposed a third way of teaching mathematics which, in the case of geometry, begins with an "intuitive" phase during which students work through practical exercises to increase their spatial awareness and to prepare them for the subsequent phases of analytical and, eventually, strictly logical reasoning. Her brochure led to a sharp, eloquent, and at times rather derogatory,

[82]Family correspondence spring and summer 1922: EA-MBL 1.1.2.

[83]Alexander Friedmann (1888–1925) Russian mathematician and physicist.

[84]Ehrenfest to Afanassjewa, 1922: EA-MBL 1.1.2.

[85]Travel documents and correspondence, 1922: EA-MBL 2.2.9 and 2.2.2:435.

[86]T. Ehrenfest-Afanassjewa, http://www.math.ru.nl/werkgroepen/gmfw/bronnen/ehrenfest2.html (Den Haag 1924).

attack by the mathematics teacher and author E. J. Dijksterhuis,[87] who furiously defended the traditional teaching methods.[88,89,90] Their fierce debate left its traces: it resulted in the establishment of a journal on the didactics of mathematics, *Euclides*, that exists until this day.[91] Also, a committee was instituted to redesign mathematics education at the HBS [Hogere Burgerschool], the secondary school that focused on the sciences.[92] It was clear from the beginning though that the time had not yet come to truly modernize the Dutch educational system: the traditionalistic Dijksterhuis was asked to become a member of the committee, while Afanassjewa with her novel methods was passed over.

She did not dwell on this defeat, as she was already busy with something else. In 1925, the *Zeitschrift für Physik* printed an article of hers that built on recent work by Constantin Carathéodory[93] and tried to give thermodynamics a firm axiomatic basis.[94] By precisely defining concepts like "reversibility" and "thermal coupling," she was able to cast the Second Law of Thermodynamics into a new light, and she also demonstrated that Kelvin's and Clausius's definitions of it are not equivalent.[95] During the rest of her life, she would—with interruptions—continue to work on her ideal of a more systematically organized thermodynamics. It would eventually result in the book *Die Grundlagen der Thermodynamik*, published in 1956 by Brill, Leiden, on her instigation and at her own expense.[96]

Afanassjewa had high hopes for this book that expanded her ideas about thermodynamics. "I believe my book relates to an introductory—and more experimentally oriented—course (in thermodynamics), as Euclid's Elements (no arrogance) relates to an introductory geometry course," she wrote to Einstein, to whom she had sent a manuscript in 1947.[97] It is difficult to say whether she considered the book equally fundamental as her 1925 article in the *Zeitschrift*. In any case, the article did not raise as much interest as she had probably hoped for at the time. Physics had been shaken up by Einstein's theory of general relativity and it was right in the middle of the quantum revolution. To most physicists, and especially to the younger ones, thermodynamics seemed to be a topic from the previous century: old-fashioned. As a result,

[87]Eduard Jan Dijksterhuis (1892–1965) was a Dutch mathematics teacher and author of several well-known books on the history of science with a focus on the role of mathematics: Klaas van Berkel (1996).

[88]Dijksterhuis (1924a–1925a).

[89]Ehrenfest-Afanassjewa (1924a–1925a).

[90]Dijksterhuis (1924b–1925b).

[91]Klomp (1997); De Moor (1999); Van Berkel (2000).

[92]Mandemakers (1996).

[93]Carathéodory (1909), (1925).

[94]Ehrenfest-Afanassjewa (1925a), (b).

[95]Uffink (2001).

[96]Ehrenfest-Afanassjewa (1956); Correspondence with Brill: EA-MBL 2.2.9:428.

[97]Ehrenfest-Afanassjewa to Einstein, 18 August 1947: The Albert Einstein Archives at the Hebrew University of Jerusalem (AEA-HUJ), doc. 10–343.

the novelty of some of Afanassjewa's insights escaped them. As so often in her life––during her studies and in the debates about mathematics education—Afanassjewa's timing was off.

Did this influence her decision to return to Russia, where people appreciated her work and were happy to invite her to teach and discuss educational reforms? Was she under the spell of the enthusiastic stories about Soviet research that were told by Russian academics, like Ioffe or Baumgart who came to visit Leiden? Between 1926 and 1933, Afanassjewa would regularly travel to her native country and stay there for six to nine months to train mathematics teachers, who were so urgently needed in the young Soviet Union and to help develop new teaching methods. She taught at Simferopol (Crimea) and Ordzhonikidze (Caucasus), and she worked in Leningrad and Moscow,[98] but the brief remarks on her postcards to family members do not reveal very much about her personal circumstances.[99]

1.8 Drifting Apart, the Downhill Slope

"You will feel embarrassed when walking next to me [again]," Afanassjewa, in Moscow, wrote to Ehrenfest, in Leiden, in 1932, "because my coat is worn threadbare."[100] Postcards to her mother regularly expose the lack of organization at the universities and polytechnic schools in the young Soviet Union. "I still do not have much work here, since the fourth-year students whom I gave a practical course about didactics, received their diploma before finishing this course. (…) I worked in good harmony this year, but I have not been able to introduce new teaching methods, since they [her students] already had traineeships at schools where teachers could not care less about such methods," she wrote, for example, during her stay in Simferopol.[101]

Yet, Afanassjewa seemed to believe in the socialist experiment taking place in her motherland. In a letter for Galinka's twenty-second birthday, written between two trips to Russia, she urged her daughter, who loved drawing and painting: "Continue to behave yourself properly, develop yourself and look for—I would really like that—a possibility to perform useful work in Russia; find yourself a like-minded gentleman and remember, in time, to also take an objective enough stance in these matters, that is: such that you will not lose your capacity to work under any circumstance! What else should I advise you? Most important is to have a goal that transcends the arts

[98]Ehrenfest-Afanassjewa (1928a, b, 1930a, b, 1931a, b). [All translated from Russian to Dutch by Hans Driessen].

[99]Klein (n.5); Letters to relatives show that she mainly worked in Simferopol between 1926 and 1930 and in Moscow, where she stayed with the Mandelstam and Kagan families, from the end of 1931 until the middle of 1932; in Ordzhonikidze she worked from the end of 1932 until the middle of 1933: EFA [translated from Russian to Dutch by Hans Driessen].

[100]Ehrenfest-Afanassjewa in Moscow, to Ehrenfest in Leiden, 28 March 1932: EFA [translated from Russian to Dutch by Hans Driessen].

[101]Ehrenfest-Afanassjewa in Simferopol to C. Afanassjewa in Leiden, 24 March 1930: EFA [translation ibidem.].

and to serve that goal through the arts: that is what makes art valuable. Illustrations for children, posters to elevate the masses, those are marvelous!"[102]

Galinka ignored her advice. Like her father had done in 1924 and 1930,[103] she traveled extensively through the United States and visited Ehrenfest's older brother Hugo, who was an influential gynecologist living in Saint Louis.[104] The family, once so close-knit, became dispersed. Ehrenfest, who was not as socialistically inclined as his wife, but was still close to his Russian friends, traveled all over Europe and sometimes into the Soviet Union as well. Vassik remained in his institution in Jena, Germany. Tanichka, the eldest daughter who was a talented and skilled mathematician, made several trips to Moscow and the Crimea and stayed in Göttingen at length, before devoting most of her time to her husband and six children.[105] Pawlik was sent to England and France to become more proficient in foreign languages.

In between such trips, Pawlik and Galinka were often home alone with baba Katya who had grown old and deaf. Aunt Sonya, who had also been living in the house in Leiden since 1916, had moved to Paris in 1929.[106] She was unhappy with Afanassjewa's visits to Russia but "I will stop complaining about Russia now," she ended a long and reproachful letter to her niece in 1931, because: "you believe these pretty stories of Bolsheviks like Ioffe, Kagan and the like anyway and do not want to hear anything bad about them."[107]

Afanassjewa and Ehrenfest were gradually growing apart, and Ehrenfest's depressions deepened.[108] He regularly withdrew into his private rooms, where he tinkered with radios and wrote long and plaintive letters, alienating himself from his friends. Fewer and fewer visitors came to the house in Leiden, especially since Ehrenfest had told quite a few of them that he did not have much more to teach them. Increasingly worried about the situation in Germany, he drew up lists of Jewish colleagues in Germany who needed help and he toyed with the idea of giving up his chair in Leiden for one of them.[109] A love affair with Nelly Posthumus-Meyjes, an art critic who was ten years his junior, destabilized him even more.[110]

Perhaps for the first time in her life, Afanassjewa did not know what to do. After Ehrenfest had reluctantly told her that he wanted a divorce, she went to the Belgian resort of Spa, completely exhausted.[111] Unexpectedly, she had "a merry time" with

[102]Afanassjewa to Galinka Ehrenfest, 12 July 1932: EFA [translation ibidem.].

[103]Trips of several months in 1923/1924 and in 1931/1932: family correspondence EA-MBL 1.1.2; Hollestelle (n.81).

[104]Ashwal (1990).

[105]Ehrenfest (1931); N.G. de Bruijn, In memoriam T. van Aardenne-Ehrenfest, 1905–1984, *Nieuw Archief voor Wiskunde 3:4* (1985) 235–236.

[106]Family correspondence 1922–1933: EA-MBL 1.1.2.

[107]Afanassjewa-Maslova to Afanassjewa, 1931: EFA [translated from Russian to Dutch by Hans Driessen].

[108]Hollestelle (n.81).

[109]Hollestelle (n.81).

[110]Van Delft (2014).

[111]Correspondence Ehrenfest-Afanassjewa (1932): EFA.

Einstein, who had come to visit her there. "Could you in less than no time send me my booklet 'Übungensammlung,'[112] because Einstein is really interested in it and he would now have time to take a peek at it," she wrote to Ehrenfest. "He is very pleased with life here [in Spa] although some people already turned out to see him yesterday, but that was a Sunday, perhaps they will leave him alone on working days."[113] Einstein, whose life was in danger in Nazi Germany, was still a hero to many people in Belgium and elsewhere. Yet, who was in the better position to help the other in these bleak and joyless times?

In 1933, when Einstein had fled Germany and was staying on the East coast of England on his way to the United States, the news reached him that Ehrenfest had died. His friend had eventually done what he had already announced, in greater or lesser detail, in letters to his friends—some of which were never sent—in the preceding months.[114] Having first shot Vassily, who had meanwhile been transferred to an institution in Amsterdam, he had then killed himself[115] with an old Browning revolver that he kept in a drawer.[116]

1.9 Hardship with Integrity

"Modest," "full of energy for the things she believed in," "pleasant," "someone with the capacity to really listen," "an unusually strong and steadfast personality and his [Ehrenfest's] intellectual equal," someone that faced "hardship" with "integrity."[117] That is how others characterized Afanassjewa's personality, or her conduct in relation to Ehrenfest. She would be in dire need of those qualities now, living as she did in the large house with only her mother and her youngest daughter. The eldest daughter, Tanichka, had moved to Dordrecht after her marriage. Pawlik was to leave for Paris soon, where he would stay and work for a long term with the physicist Pierre Auger.[118] Afanassjewa's visits to the Soviet Union were a thing of the past: having witnessed the Holodomor in Ukraine[119] in 1932 and having heard about Stalin's atrocities, she had decided not to return to the Soviet Union anymore.[120]

---

[112] T. Ehrenfest-Afanassjewa (1931). The booklet contains exercises to increase the spatial awareness of students during a preparatory phase in geometry teaching.

[113] Afanassjewa to Ehrenfest, 15 August 1932: EFA [translated from Russian to Dutch by Hans Driessen].

[114] E.g., Pais (1991); Van Delft (n.110).

[115] E.g., Hollestelle (n.81).

[116] Personal communication T. van Bommel.

[117] Van Hiele and Krooshof (1964a); Einstein (1934), [reprinted] in: Einstein (1950); Burgers (n.12) 51; personal communication H. Langedijk.

[118] Pierre Auger (1899–1993) a French physicist who conducted cosmic ray experiments in which Pawlik was taking part.

[119] The man-made famine in Ukraine, 1932–1933, in which millions of Ukrainians died.

[120] In 1932, Ehrenfest's former student Jan Burgers gave up his membership of the Dutch communist party, based on her advice. Personal communication Herman Burgers.

More sorrow was to follow. In January 1939, Pawlik was killed in an avalanche in the French Alps, doing what he liked best: skiing. World War II broke out. In 1943, her son-in-law, the painter and children's book illustrator Jaap Kloot,[121] who had married Galinka in 1941, was deported and killed in Sobibor. Galinka lost their unborn child after she had been imprisoned in Den Bosch for a week.[122]

Afanassjewa herself went back to teaching for a while. She taught *Foundations of Physics and Mathematics* as a "privaatdocent," a private teaching position associated with Leiden University. She worked at the university from March 1941 till June 1942, when she resigned in protest because regulations to ban Jewish professors and teachers were applied ever more strictly.[123] However, her decision, in 1941, to teach at the university at all, especially with her background, raised some eyebrows. It may have been a strategic choice, made to protect Galinka and perhaps in order not to draw undue attention to other people who were temporarily living in hiding at Witte Rozenstraat 57 during the war years.[124]

Studying and writing were her lifeline during the remainder of the war and shortly thereafter. It helped her through the days when the walls in the large house appeared to close in on her, seemingly resonating with the laughter and voices from the past and keeping her from falling asleep any earlier than five in the morning, as her granddaughter remembers.[125] It also resulted, among other things, in an essay entitled *Relevia a new economic system, an order in which I myself would like to live.* The booklet described a state with a government that was a mix of socialism and

[121] Jacob Kloot (1916–1943) married Galinka Ehrenfest in 1941. Both wrote and illustrated children books under the name El Pintor, which were published by Kloot's publishing house Corunda in Amsterdam. See, e.g., RKD Dutch Institute for Art History: *rkd.nl/en/explore/artists/111285.* In daily life, Jacob and other family members used "Kloots" as their family name.

[122] Galinka Ehrenfest was arrested while visiting new addresses to hide Jewish relatives and friends. She was released after one week in prison on 26 June 1943. Personal communication T. Van Bommel and documents from the German *Sicherheitspolizei* in EFA. After the war she married Henk van Bommel [n.13] and had a daughter with him.

[123] Afanassjewa requested to be allowed to teach the basic principles of mathematics and physics as a *privaatdocent*, associated with the University of Leiden in November 1940, shortly before the dismissal of Jewish professors and the famous protest speech of Prof. Dr. Cleveringa (26 November 1940). After this speech, the university was closed temporarily at first and, when the German authorities were unable to subject the university to Nazification, definitely in November 1942. Afanassjewa was granted the position on March 10 1941. She requested that the authorities terminate her contract on 11 May 1942. Her request was granted on 15 June 1942. The Secretary General of the Department of Education, Science, and Protection of Culture to Afanassjewa, 10 March 1941, accepting her as *privaatdocent*; Ibidem.15 June 1942, terminating her contract at her request: EA-MBL 2.5.

[124] Two sisters of Jaap Kloot stayed there temporarily: personal communication T. van Bommel. The elder brother of H. Langedijk stayed there at the end of the war, in order to escape the *Arbeitseinsatz* (forced labor and slavery in Nazi Germany) and during the week when H. Langedijk visited him, close to the end of the war, she counted at least four other people in hiding, among whom was the Jewish Mrs. Pinto: Personal communication. A certificate and letter testify that a tree was planted in Afanassjewa's honor in the "Joop Westerweel-woud," Ramat Menashe, Israel, by Mr. and Mrs. Schaap-Redak in 1946: EA-MBL, 2.2.32:453.

[125] Personal communication.

liberalism,[126] which Afanassjewa had often discussed with Galinka, as well as with Ehrenfest's former student Jan Tinbergen who used his training in physics to develop the new field of econometrics and who would eventually be awarded a Nobel prize (in 1969) for his efforts.[127]

Another work that Afanassjewa wanted to publish was her book: *Die Grundlagen der Thermodynamik*. She turned to Jan Burgers, who was a professor at the Laboratory for Aerodynamics and Hydrodynamics of Delft University and also a former student of Ehrenfest. At her request, Burgers passed on the manuscript to Erwin Schrödinger in Dublin, who had just finished a short paper on statistical thermodynamics himself.[128] In the meantime, Afanassjewa had sent the manuscript to Einstein, in the hope that he would take it upon himself to find her an American publisher because, as she wrote to him, "in Europe there is not enough paper… and even fewer big shots who can recommend my book to a publisher."[129]

Einstein had stayed in touch with the Ehrenfest family. Immediately after the war, he had asked a friend of Helen Dukas, his secretary, to visit Leiden and to find out whether the family members were all right.[130] For the next year and a half, he regularly sent Afanassjewa food parcels,[131] until she asked him to send them to Edmund Bauer in Paris instead because the Parisians suffered from hunger more than she did in Leiden.[132] In the summer of 1947, Einstein again proved himself to be "really a true friend,"[133] Afanassjewa wrote, because he immediately began reading her manuscript.

She was not nearly as pleased with Einstein's opinion once he had read the entire manuscript. Her ideas seemed interesting and sound to him, and he thought that the book should be published ("in English!"), he wrote. Yet, the text in its current form was too long, and difficult to read. It had made him feel "a little bit" like a spectator at "a magician's show with so many nice details to look at that one does not notice when the frog of your unforgettable P.E. [Paul Ehrenfest] jumps into the water."[134] The frog was a reference to a well-known phrase that Ehrenfest had often used to indicate the essence of a theory, a hypothesis, or a line of reasoning.

---

[126]Ehrenfest-Afanassjewa (1946).

[127]Although Tinbergen did not necessarily agree with all Afanassjewa's views, he arranged for the work, which was published in 1946 at Boucher, The Hague, to be reviewed by the economist Prof. Dr. H.J. Seraphim in *Weltwirtschaftliches Archiv*: Tinbergen to Ehrenfest-Afanassjewa, 16 July 1955: EA-MBL 2.2.9: 482.

[128]Schrödinger to Burgers, 22 March 1946: EA-MBL 2.2.9:482. In the following years, Schrödinger would summarize his ideas in Schrödinger (1952).

[129]Afanassjewa to Einstein, 17 August 1947: (AEA-HUJ), doc. 10–312.

[130]Dukas to Afanassjewa, 17 April 1946: EA-MBL 2.2.1: 418; Prue Smedts to Afanassjewa, 30 September 1945: EA-MBL 2.2.1:418.

[131]The parcels were prepared by Mimosa Food Parcels in New York: EA-MBL 2.1.

[132]Einstein to Afanassjewa, 28 March 1947: EA-MBL 2.1. Einstein comments on her request.

[133]Afanassjewa to Einstein, 17 August 1947 (n.129).

[134]Einstein to Afanassjewa, 12 August 1947: EA-MBL 2.2.9: 482.

It had been a deliberate choice by Afanassjewa to ignore both the frog and her own ideas about didactics.[135] Instead of beginning with a more intuitive exploration and then working toward a formal and axiomatic system, she had immediately plunged into definitions, axioms and formal, logical reasoning. Yet, according to Einstein, she gave the impression that "a logical cleaning devil [Putzteufel]" had taken possession of her.[136] He added affably that he might not be authoritative in this case, since "I am at the other extreme, dirty and economical with cleanliness."[137] Still, his conclusion remained: Afanassjewa's wish to tidy up the field of thermodynamics and describe every concept and detail in a way that was too orderly and precise had rendered the text a bit obscure. Moreover, Einstein did not know anyone who could be entrusted with the translation of the text.

Afanassjewa protested to no avail. The "creative" Einstein was perhaps not inclined to "cleaning devilishly," she wrote, but "you did admit, that you learned something from my presentation [of thermodynamics]: well, for this we are indebted to the cleaning devil."[138] She continued ironically: "That you wash your hands of my book, surely is very sad: I was hoping I could die in peace! What shall I do now?", but Einstein was too busy to reply.

Did Einstein realize how intensely Afanassjewa was invested in her work? An ocean, the Second World War, and many years separated his European life from his current American existence. How happy he had been in Leiden: lunch with Afanassjewa and her aunt Sonya, playing music with Ehrenfest, laughing and playing with the children, walks through the streets of Leiden or along the beach at Noordwijk, discussions on anything from politics to literature to physics. "Yes, that was really a beautiful and tranquil time that we spent together," he had written after one such visit in 1919[139]—and many more joyful gatherings had taken place before Ehrenfest sank into, and eventually drowned in, his deep depressions.

In November 1948, Afanassjewa picked up the thread of their friendship and sent Einstein a handwritten letter: "Fortunately, I can look at your picture in some newspaper now and then." She continued by describing how her daughters were doing and also referred to Ehrenfest. About herself she remarked only: "I feel more fit now than two years ago: the years of hunger, cold—and worries—apparently had a longer lasting influence."[140] Without further ado, they resumed their correspondence, writing about physics, mathematics, and family. "Liebe Frau Ehrenfest," Einstein replied, in April 1953, to a letter in which Afanassjewa apologizes for having forgotten his birthday.[141] "You are absolutely right not to burden your brain with dates

[135]See also (n.97): "I believe my book relates to an introductory—and more experimentally oriented—course (in thermodynamics), as Euclid's Elements (no arrogance) relates to an introductory geometry."

[136]Einstein to Afanassjewa (n.134).

[137]Ibid.

[138]Afanassjewa to Einstein (n.97).

[139]Einstein to Ehrenfest, 9 November 1919: CPAE 9, Doc. 155. Also cited in Klein (n.5) 313.

[140]Afanassjewa to Einstein, 28 December 1948: (AEA-HUJ) Doc.10–314.

[141]Einstein to Afanassjewa, 18 April 1953: EA-MBL 2.1.

of birth. I receive your congratulations equally enthusiastically, no matter what day of the year you pick." It was to be one of his last letters to her.

1.10 Final Years in Her Own Sphere

Afanassjewa's days were spent quietly now. To provide herself with an income, she rented out the rooms on the first and the attic floor of her large house, she gave private Russian lessons, and she taught a few courses at Leiden University.[142] She worked in her garden, carried on a lively correspondence with her many friends abroad, and continued to participate in debates about the didactics of physics and, especially, mathematics.

After Ehrenfest's death, her role in the community of mathematics educators had become increasingly important. In 1936, three years after her husband died, she had been asked to join the Dutch *Wiskunde Werkgroep*, a discussion group that met, almost always at her house, to discuss possible educational reforms in The Netherlands.[143] In 1938, she had been the central figure at a long weekend meeting of this Working Group, presiding over the debates and introducing all speakers. After World War II, the influential mathematician and pedagogue Hans Freudenthal,[144] who had called her *Übungensammlung* a "masterpiece" in the field of didactics of mathematics,[145] had become so deeply inspired by her work that—in retrospect—it is not always easy to see where her ideas end and his ideas begin.

Her work on thermodynamics and statistical mechanics drew less attention in her later years. When *Die Grundlage der Thermodynamik* was eventually printed, in 1956, she sent copies of the book to many acquaintances, but their reactions were, at best, polite and certainly not enthusiastic.[146] Three years later, in 1959, Cornell University Press published the well-known *Enzyklopädie* article on statistical mechanics by the two Ehrenfests in an English translation.[147] Giving most of the credit for this work to her husband, Afanassjewa herself writes in the preface to the translation: "The great task of collecting literature and of organizing the Enzyklopädie article was done by Paul Ehrenfest." [...] "My contribution consisted only in discussing with him all the problems involved, and I feel that I succeeded in clarifying some concepts that were often incorrectly used." These modest words were grist to the mill of those who wanted to belittle her contribution to their collaborative efforts.

---

[142]Lecture notes of her course on thermodynamics 1947/1948 are available in the Gorlaeus Library, Leiden University.

[143]De Moor (1999).

[144]Hans Freudenthal (1905–1990) was a professor of mathematics at Utrecht University, who later specialized in the didactics of mathematics. The Freudenthal Institute for Science and Mathematics Education at Utrecht University is named after him. See also, e.g., Bastide-van Gemert (2006); Molenaar (1994).

[145]De Moor (1993).

[146]Correspondence: EA-MBL: 2.2.9:482.

[147]Paul and Tatiana Ehrenfest (1959).

In their preface, Uhlenbeck and Kac take this division of labor for granted and write that "The late Paul Ehrenfest" prepared the article "*in collaboration with his wife.*" In his Ehrenfest biography, in 1970, Klein is equally neglectful of Afanassjewa's work on statistical mechanics, stating only that "Paul Ehrenfest was primarily responsible for this article (...) as Tatiana herself wrote almost half a century later."[148,149]

Another short paper, on the notion of probability, appeared in the *American Journal of Physics* in 1958. It was an adapted translation of a paper Afanassjewa had published earlier in Russia, in 1911. At the time, it had served as an interesting contribution to the debate,[150] but four decades later it did not add much that was new. Nevertheless, these publications must have pleased Afanassjewa. They are enduring proof that she had been a full-fledged scientist and not just a professor's wife who had stayed in the background and kindly received her husband's colleagues while taking care of the children. Similarly, it must have pleased her that, in 1961, the Dutch mathematician Bruno Ernst took the initiative to put together a compilation of her essays on the didactics of mathematics and publish them with a preface describing her career.[151]

And so, Tatiana spent her final years in her own sphere, in a country that she never fully came to appreciate, often together with what remained of her family. Afanassjewa's granddaughter still remembers how happy "baba Tanja" was when she came to visit Galinka, who now lived in a small village in Limburg, as it was one of the few Dutch provinces with hills and woods that vaguely resembled the Russian forests.[152] Afanassjewa would never learn to like the flat polders surrounding Leiden, or the city's reserved inhabitants. In her own large house, she managed, until the very end of her life, to maintain a somewhat Russian atmosphere, both with regard to hospitality and intellectual challenges. "Until she died, her house at Witte Rozenstraat 57 in Leiden has been a place where mathematics and physics teachers went to verify and discuss new didactical insights," her former colleagues wrote after her death.[153] "They never went home empty-handed."[154]

Acknowledgements I would like to particularly thank Dr. Henriette Schatz for carefully correcting mistakes in the text and for her helpful suggestions; Prof. Dr. Jos Uffink for extremely valuable discussions about, among other things, Afanassjewa and thermodynamics; Prof. Dr. Anne Kox for pleasant and encouraging conversations throughout the project, Prof. Dr. Diana Kormos-Buchwald for her help in locating letters from Afanassjewa to Einstein, the late Dr. Ed de Moor for helpful conversations about Afanassjewa and the didactics of mathematics, Tamara van Bommel for opening up the family archive and Dr. Hans Driessen for carefully translating the many Russian texts. Finally, I would like to thank Stichting Physica for a grant (2,200 euros) for translating these Russian texts.

---

[148] Klein (n.5) 121.

[149] Ehrenfest & Ehrenfest (n.147) 7–9.

[150] Ehrenfest-Afanassjewa (1958, 1911a, b, c)..

[151] Ehrenfest-Afanassjewa (n.17).

[152] Private communication.

[153] Afanassjewa died at her home in Leiden on 14 April 1964.

[154] Van Hiele G. Krooshof (1964a).

References

Ashwal, S. (Ed.). (1990). *The founders of child neurology* (483–495). San Francisco.

Bastide-van Gemert, S. (2006). *"Elke positieve actie begint met critiek": Hans Freudenthal en de didactiek van de wiskunde*. Ph.D. Dissertation University of Groningen.

Blaauboer, M. (2015). Hendrika J. van Leeuwen. Portret van de eerste vrouwelijke lector bij Technische Natuurkunde in Delft. *Nederlands Tijdschrift voor Natuurkunde, 81*, 4–6.

Bonner, T. (1995). *To the ends of the earth: Women's search for education in Medicine*. Cambridge.

Carathéodory, C. (1909). Untersuchung über die Grundlagen der Thermodynamik. *Mathematische Annalen, 67*, 355–386.

Carathéodory, C. (1925). Über die Bestimmung der Energie und der absoluten Temperatur mit Hilfe von reversiblen Prozessen. In *Sitzungsberichten der Preussischen Akademie der Wissenschaften Physikalisch-Mathematische Klasse* (pp. 39–47).

De Haan, F. (1998). *Gender and the politics of office work: The Netherlands 1860–1940* (pp. 122). Amsterdam.

De Haas-Lorentz, G. L. (1913). *Die Brownsche Bewegung und Einige Verwandte Erscheinungen*. Braunschweig.

De Moor, E. (1993). Het gelijk van Tatiana Ehrenfest-Afanassjewa. *NW Tijdschrift voor wiskunde, 12*(4), 18–25.

De Moor, E. (1999). *Van vormleer naar realistische meetkunde*. Ph.D. Dissertation University of Utrecht.

Dijksterhuis, E. J. (1924a–1925a). Moet het meetkunde-onderwijs gewijzigd worden? Opmerkingen n.a.v. een brochure van mevrouw Ehrenfest-Afanassjewa. *Bijvoegsel op het Nieuw Tijdschrift voor Wiskunde, I*, 1–26).

Dijksterhuis, E. J. (1924b-1925b). Antwoord op mevrouw Ehrenfest-Afanassjewa. *Bijvoegsel op het Nieuw Tijdschrift voor Wiskunde, I*, 60–68.

Ehrenfest, T. (1931). *Oppervlakken met scharen van gesloten geodetische lijnen*. Ph.D. Dissertation University of Leiden.

Ehrenfest, P., & Ehrenfest, T. (1959). *The Conceptual Foundations of the Statistical Approach in Mechanics*. Ithaca.

Ehrenfest-Afanassjewa, T. (1911a). On the application of probability theory to natural phenomena. *Journal of the Russian Physicochemical Society/JETP, 5*.

Ehrenfest-Afanassjewa, T. (1911b). The principle of conformity and its applications. *Journal of the Russian Physicochemical Society/JETP, 7*.

Ehrenfest-Afanassjewa, T. (1911c). Over de toepassing van de waarschijnlijkheidstheorie op wetmatige verschijnselen. *Tijdschrift van het Natuur-en Scheikundig Genootschap, sectie Natuurkunde, 5*.

Ehrenfest-Afanassjewa, T. (1912). The principle of dimensions. *Journal of the Russian Physicochemical Society/JETP, 7*.

Ehrenfest-Afanassjewa, T. (1914). On the principle of corresponding states *Journal of the Russian Physicochemical Society/JETP, 7*.

Ehrenfest-Afanassjewa, T. (1924a–1925a). Moet het meetkunde-onderwijs gewijzigd worden? Een antwoord aan de heer E. J. Dijksterhuis. *Bijvoegsel op het Nieuw Tijdschrift voor Wiskunde I*, 47–59.

Ehrenfest-Afanassjewa, T. (1925a). Zur Axiomatisierung des zweiten Hauptsatzes der Thermodynamik. *Zeitschrift für Physik, 33*, 933–945.

Ehrenfest-Afanassjewa, T. (1925b). Zur Axiomatisierung des zweiten Hauptsatzes der Thermodynamik. *Zeitschrift für Physik, 34*, 638.

Ehrenfest-Afanassjewa, T. (1928a). Irreversibility and the second law of thermodynamics. *[Russian] Journal of applied physics*, 3–4.

Ehrenfest-Afanassjewa, T. (1928b). Spatial intuition and physical experience. *Communications of the Pedagogical Academy M. V. Froense, 2*, 5.

Ehrenfest-Afanassjewa, T. (1930a). How to begin teaching geometry? *Physics, Chemistry and Engineering at the Soviet Schools, 5.*
Ehrenfest-Afanassjewa, T. (1930b). On the interpretation of the second law of thermodynamics in the theory of Max Planck. *(Russian) Journal of Applied Physics, 1.*
Ehrenfest-Afanassjewa, T. (1931a). The results of kinetic theory. *[Russian] Journal for Experimental and Theoretical Physics, 6.*
Ehrenfest-Afanassjewa, T. (1931b). *Uebungensammlung zu einer geometrischen Propädeuse.* Den Haag.
Ehrenfest-Afanassjewa, T. (1946). *Relevia: een nieuw economisch systeem, een orde, waarin ik zelf ook graag zou willen leven.* Den Haag.
Ehrenfest-Afanassjewa, T. (1956). *Grundlagen der Thermodynamik.* Leiden.
Ehrenfest-Afanassjewa, T. (1958). On the Use of the Notion "Probability" in Physics. *American Journal of Physics, 26,* 388.
Ehrenfest-Afanassjewa, T. (1960). *Didactische opstellen wiskunde.* Zutphen.
Einstein, A. (1934). In memoriam Paul Ehrenfest.
Einstein, A. (1950). *Out of my later years* (214–217). New York.
Frenkel, V. J. (1977). *Paul Ehrenfest.* Moscow.
Hans, N. (1963). *The Russian tradition in education* (pp. 72–73). New York
Hollestelle, M. J. (2011). *Paul Ehrenfest, Worstelingen met de moderne wetenschap.* Ph.D. Dissertation University of Utrecht.
Klein, M. J.(1970) *Paul Ehrenfest: The making of a theoretical physicist.* Amsterdam.
Klomp, H. (1997). *De relativiteitstheorie in Nederland.* Ph.D. Dissertation University of Utrecht.
Koblitz, A. (2013). *Science, women and revolution in Russia* (p. 26). New York.
Kox, A. J. (Ed.). (2018). *The scientific correspondence of H. A. Lorentz, the Dutch correspondents,* Vol. 2. New York.
Mandemakers, C. A. (1996). *Gymnasiaal en middelbaar onderwijs: ontwikkeling, structuur, sociale achtergrond en schoolprestaties, Nederland, ca. 1800–1968.* Ph.D. Dissertation, Erasmus University Rotterdam.
Mary, R. S. (2015). Creese, *ladies in the laboratory IV: Imperial Russia's women in science* (pp. 92). London.
Molenaar, L. (1994). *"Wij kunnen het niet langer aan de politici overlaten..."–De geschiedenis van het Verbond van Wetenschappelijke Onderzoekers (VWO), 1946–1980* (pp. 159–162). Delft.
Morrissey, S. K. (1998). *Heralds of revolution, Russian students and the mythologies of radicalism.* Oxford.
Pais, A. (1991). Niels Bohr's Times. In *Physics, Philosophy, and Polity.* Oxford.
Popović, M. (2003). In *Albert's shadow, the life and letters of Mileva Marić, Einstein's first wife.* Baltimore.
Schrödinger. (1952). *Statistical Mechanics.* Cambridge.
Senger, J. V., & Ooms, G. (Eds.) (2007) Autobiographical notes of Burgers, *J. M. Burgers centre at 15 years* (p. 50). Delft.
Steen, A. V. (2011). Vol moed en blakende ijver—Aletta Lorentz-Kaiser en de vrouwenbeweging in Leiden (1881–1912). *Jaarboek Dirk van Eck.* Leiden.
Stites, R. (1978). *The women's liberation movement in Russia: Feminism, Nihilism, and Bolshevism 1860–1930* (p. 83). Princeton.
Thiele, R. (2011). *Felix Klein in Leipzig.* Leipzig.
Tsipenyuk, Yu. (1973). From the history of the journal of the Russian physicochemical society/JETP. *Soviet Physics-Journal of Experimental and Theoretical Physics, 37*(1), 1–24.
Uffink, J. (2001). Bluff your way into the second law of thermodynamics. *Studies in History and Philosophy of Science Part B: Studies in History and Philosophy of Modern Physics, 32*(3), 305–394.
van Berkel, Klaas. (2000). De geboorte van een tijdschrift. *Euclides, 75*(4), 111–116.
van Delft, Dirk. (2014). Paul Ehrenfest's final years. *Physics Today, 67*(1), 41–47.
Van der Heijden, M. (2015). Schoolbord van de moderne fysica, *NRC Handelsblad.*

van Hiele, P. M., & Krooshof, G. (1964). Tatiana Ehrenfest Afanassjewa. *Euclides, 39*(9), 257–259.
Weigand, H. G., et al. (2017). What is and what might be the legacy of Felix Klein? In *Proceedings of the 13th international congress on mathematical education*. Cham.

Chapter 2
Intuition, Understanding, and Proof: Tatiana Afanassjewa on Teaching and Learning Geometry

Marianna Antonutti Marfori

2.1 Introduction

Tatiana Alexeyevna Afanassjewa (from 1904 Tatiana Ehrenfest-Afanassjewa) was a mathematician, a physicist, and a teacher. All three of these vocations come together in her philosophy of geometry, which bases a novel approach to the teaching of geometry on her understanding of the proper roles of intuition and logical reasoning in geometry, grounded in our experience of concrete objects occupying physical space. Having been a student at Göttingen during the time of its greatest flowering as a centre of mathematical research, and a member of the physics community during the revolutionary period from Einstein's *annus mirabilis* of 1905, she was close to the centre of some of the most exciting developments in science of her time. Since early on she was also deeply invested in teaching, and in developing new and better ways to communicate her subject to her students. Afanassjewa's reflections on the teaching of geometry are thus those of a mathematician and a theoretical physicist who was passionate about scientific discussion and teaching: her ideas originate in her own experience as a student, researcher, and teacher, and in the debates with her scientific contemporaries—debates in which she played an active and important role.

The writing of Sects. 2.1–2.3 of this paper was supported by funding from the European Union's Horizon 2020 research and innovation programme under the Marie Skłodowska-Curie grant agreement No. 709265. The writing of Sects. 2.3–2.6 was supported by the German Research Foundation (Gefördert durch die Deutsche Forschungsgemeinschaft (DFG)—Projektnummer 390218268). Their support is gratefully acknowledged. My thanks to Benedict Eastaugh and two anonymous referees for comments on a previous draft of this paper, and to Paolo Bussotti for helpful discussions about the history of geometry. I also thank the ETH Zurich University Archives for making available to me the correspondence between Tatiana Afanassjewa and Paul Bernays. The visit to the ETH Zurich University Archives was funded by LMUexcellent within the framework of the German Excellence Strategy.

M. Antonutti Marfori (✉)
Ludwig-Maximilians-Universität München, Munich, Germany
e-mail: Marianna.AntonuttiMarfori@lrz.uni-muenchen.de

J. Uffink et al. (eds.), *The Legacy of Tatjana Afanassjewa*, Women in the History of Philosophy and Sciences 7, https://doi.org/10.1007/978-3-030-47971-8_2

The focus of this paper is Afanassjewa's philosophy of geometry and geometry education as espoused in her 1924 manifesto *Wat kan en moet het Meetkunde-onderwijs aan een niet-wiskundige geven?* (*What can and should geometry education offer a non-mathematician?*), Ehrenfest-Afanassjewa (1924a), translated into English in this volume for the first time, and in her 1931 booklet of exercises for an introductory geometry course *Übungensammlung zu einer geometrischen Propädeuse* (*Exercises in Experimental Geometry*), Ehrenfest-Afanassjewa (1931), translated into English by Hoechsmann (2011).[1] The aim of this paper is to present her ideas on geometry from a philosophical perspective, particularly her account of the roles of intuition and logic in the teaching and learning of geometry, and the relations between her ideas and those of the most prominent members of the Göttingen school, Felix Klein and David Hilbert. It will also briefly explore how her ideas about the teaching of geometry were received in the Netherlands, where she was to spend most of her life. Her extensive international training and experience, her broad intellectual culture, her curiosity, and her determination to go beyond conservative dogmas of mathematical education led her to bring a radically novel approach to the teaching of geometry in the Netherlands.

Afanassjewa's philosophical views about intuition and logical thinking are also similar in spirit to the pragmatist tradition, especially in the close connection she posits between knowledge and understanding (and it is surely the case that she was acquainted with Poincaré's work, who is explicitly mentioned in Ehrenfest-Afanassjewa 1924a). However, for reasons of space, these philosophical aspects of her thought will not be analysed in this paper.[2]

Naturally, her studies in Saint Petersburg before she went to Göttingen, as well as her engagement with the contemporary debate on mathematics education in Russia also influenced her profoundly. Regrettably, this paper cannot engage with Afanassjewa's writings in Russian, nor with Russian-language sources in general. Moreover, due to a lack of detailed, accessible historical evidence—all the courses she followed, who her main interlocutors in mathematics in the first three decades of the century were, which books she had the opportunity to read, and when—it is difficult to determine in precisely what ways the Göttingen mathematicians and other important figures such as Poincaré, Pasch, and von Helmholtz influenced her work and thought.[3] Nevertheless, it will be argued below that her approach to geometry has much in common with those of Klein and Hilbert. The following section there-

---

[1]Unless otherwise noted, all page numbers given for the manifesto refer to the translation in the present volume, and all page numbers given for the *Übungensammlung* refer to the translation by Hoechsmann (2011).

[2]In this context, it would also be interesting to analyse the similarities and the differences between Afanassjewa's view and that of Ferdinand Gonseth. It is plausible to think that she was acquainted with Gonseth's work, as he is mentioned in the correspondence between Afanassjewa and Bernays in the early 1950s. However, since the first philosophical work of Gonseth dates to 1926, 2 years after the publication of Afanassjewa's manifesto, such an analysis is beyond the scope of this paper.

[3]It is surprising that there is no correspondence between Klein and Afanassjewa, since Klein in particular was an avid correspondent (for an idea of the volume of Klein's correspondence, see e.g. Schlimm 2013, p. 184). Neither the Göttingen University Library nor the Rijksmuseum Boerhaave, which holds the Ehrenfest–Afanassjewa archive, hold letters between her and other Göttingen math-

fore briefly describes some salient aspects of Klein and Hilbert's work and interests, namely, their work in geometry, their attitudes towards spatial intuition, and particularly in the case of Klein, how these interacted with their views on the teaching of mathematics.

2.2 Göttingen

When Afanassjewa went to Göttingen in 1902, it must have seemed like the centre of the world. Felix Klein was one of the world's foremost mathematicians, "a Zeus, enthroned above the other Olympians" as the physicist Max Born would later describe him (Born and Born 1969, p. 16). David Hilbert, already famous in 1895 when Klein brought him to Göttingen, had become even more well known after his 1900 address to the International Congress of Mathematicians in Paris, in which he presented his influential list of 23 mathematical problems. Moreover, both Klein and Hilbert had done seminal work in geometry. In 1872, Klein had published his *Erlanger Programm* (Klein 1872), in association with his appointment as a professor in Erlangen, in which he outlined a radical new approach to geometry that involved understanding a given geometry as characterised by the group of transformations under which it remained invariant. Hilbert's work in geometry involved a similarly grand vision: his *Grundlagen der Geometrie* (*Foundations of Geometry*) (Hilbert 1899) aimed to provide a completely axiomatic treatment of Euclidean geometry, including the mutual independence and joint consistency of the axioms.

Both the *Erlanger Programm* and the *Grundlagen der Geometrie* were fruits of a nineteenth-century revolution in geometry. For more than 2,000 years, the dominant form of geometry had been that prescribed in Euclid's *Elements*. While the status of the parallel postulate had been debated since antiquity, there seemed little question that the postulate was true; the issue was more the way in which it was to be justified. All this changed with the work of Bolyai, Lobachevsky, and Gauss in the early part of the nineteenth century, who showed that the denial of the parallel postulate did not lead to a contradiction, but instead to new geometries beyond than the familiar Euclidean one. The work of Riemann, Helmholtz, and Beltrami in the 1860s served to convince mathematicians of the acceptability of non-Euclidean geometry (Gray 1992, p. 38), and a proliferation of different geometries followed. Klein's *Erlanger Programm* was intended as a way to unify these diverse geometries, bringing order to what had become a messy discipline (Rowe 1992, p. 47). It accomplished this by

ematicians on topics related to her philosophy of geometry (though there is some correspondence with Klein about the encyclopaedia entry on the foundations of statistical mechanics which she authored together with her husband Paul Ehrenfest). The Kalliope Verbundkatalog Nachlässe only lists correspondence between Afanassjewa and the physicist Gustav Herglotz. There is, however, extant correspondence between Afanassjewa and Paul Bernays from the late 1940s and early 1950s, mostly about theoretical physics, which can be accessed in the Bernays Nachlass at the ETH Zurich University Archives. For a detailed list of courses of Klein's that Afanassjewa attended while in Göttingen, see the recently published Tobies (2020).

providing a general scheme for classifying geometries in terms of their transformation groups; two geometries were to be considered the same if their transformation groups (that is, the set of transformations of the underlying space) were isomorphic to one another. This also gave a way of separating the essential properties of a geometry from the inessential ones: the essential properties were those which were invariant under transformations.

Three other aspects of Klein's views are notable in the present context: his intuitive approach to mathematics, his interest in mathematics education, and his positive attitude towards women mathematicians. In a marked divergence from many of his peers at the time, Klein was a strong supporter of women in mathematics, believing that they should have equal educational opportunities (Tobies 2019, p. 16), and pushing for their admission to university courses (Tobies 2012, pp. 127–128; Tobies 2020). Both Grace Chisholm, the first woman to obtain a Ph.D. in Germany, and Mary F. Winston, the first American woman to obtain a Ph.D. in mathematics, obtained their doctorates under Klein. His views were shared by Hilbert, who had a number of female doctoral students and urged his colleagues to allow them to be admitted to the university "for the sake of mathematics" (Tobies 2000, pp. 31–32). It was also Hilbert who brought Emmy Noether to Göttingen and tried to have her admitted as a *Privatdozent*: he is supposed to have protested, in the face of objections from other members of the academic senate, that "This is a university, not a bathing establishment!". Mathematics at Göttingen in the 1900s therefore provided an environment which was, at least for a German university of its time, accepting of women and foreigners such as Afanassjewa.

Klein was a dedicated teacher, known for holding long meetings with his numerous doctoral students as well as with his other collaborators (Tobies 2019, p. 10). From the 1890s onwards, he also played an important role in the reform of mathematical education in Germany and internationally (see e.g. Nabonnand 2007; Furinghetti et al. 2013; Tobies 2019). Mathematics education was the theme of his inaugural address at Erlangen in 1872 (Rowe 1983, 1985) and became an increasingly important part of his work during his time in Göttingen. Klein felt that there was a lack of widespread knowledge of mathematics in society. For Klein mathematics was a formal educational tool for training the mind, and he claimed that mathematics lessons at school were not "developing a proper feeling for mathematical operations or promoting a lively, intuitive grasp of geometry" (*Erlanger Antrittsrede*, English translation in Rowe 1985, p. 139). Klein therefore worked to improve mathematics education in universities and technical schools, as well as in secondary schools: the Meran reform which took effect from 1905 onwards is sometimes referred to in German as the "Klein'sche Reform".

The "intuitive grasp of geometry" alluded to in his *Erlanger Antrittsrede* was a central feature of Klein's approach to mathematics. A well-known passage of Poincaré presents some of Klein's work on Riemann surfaces, in which Klein models such surfaces by metallic surfaces with varied electrical conductivity, as a paradigm of the intuitive approach to geometry. Poincaré comments that "Klein well knows he has given here only a sketch; nevertheless he has not hesitated to publish it; and he would probably believe he finds in it, if not a rigorous demonstration, at least a kind of moral certainty" (Poincaré 1913, p. 211). Klein placed an emphasis on intuition in his teaching and writings, and especially on the use of "Raumanschauung" or spatial intuition in geometry (Mattheis 2019). In the second volume of his textbook *Elementarmathematik vom höheren Standpunkte Aus* (*Elementary Mathematics from a Higher Standpoint*), first published in 1908, Klein appeals to spatial intuition both as giving content to the axioms of geometry and as an important method for students to acquire geometrical knowledge (Klein 2016).

Spatial intuition comes together with teaching in Klein's development and use of physical models to represent geometric objects and structures. During his time at the Technische Hochschule in Munich from 1875 to 1880, he founded a laboratory to construct such models together with his colleague Alexander von Brill. These models, commercially produced by Brill's brother, later became widely used in mathematics departments across Europe and the United States (see Rowe (1989, p. 191) as well as Bartolini Bussi, Taimina, and Isoda (2010, p. 21)). Klein viewed such models as valuable for teaching but also, following his teacher Julius Plücker, as tools for research (Rowe 2013). In Göttingen during the period that Afanassjewa was there, the library contained a large collection of such models, and in her exchange with Dijksterhuis, Afanassjewa explicitly notes her appreciation for Klein's use of wire and plaster models of spatial curves and surfaces during his lectures (Ehrenfest-Afanassjewa 1924b, pp. 48–49, fn. 3). Under Klein's guidance, the Göttingen mathematics library was developed as a *Präsenzbibliothek*: a reference library whose open shelves were packed with journal offprints, as well as volumes containing the texts of lectures by Dirichlet, Riemann, Hilbert, Klein, and others. Klein's seminar notes alone, spanning almost the entirety of contemporary mathematics, filled 29 volumes of more than 8,000 pages (Chislenko and Tschinkel 2007). The library was designed so as to facilitate not just individual research, but also to encourage informal meetings between mathematicians (Rowe 1989, p. 202). Klein himself was known to sit for hours discussing with his students and collaborators.

The sea change that took place in mathematics during the nineteenth century had another aspect, namely, the new importance attached to rigour. One manifestation of this development was the so-called arithmetization of analysis: the refounding of the calculus on the basis of set-theoretic constructions, together with rigorous definitions of such notions as limit and continuity (see e.g. Grattan-Guinness 2000, Sect. 2.7; Giaquinto 2002, Chap. 1 and Ferreirós 2007). Some arithmetizers explicitly made it their aim to displace geometric intuition, at least in a justificatory capacity, in favour of principles they regarded as more rigorous. Richard Dedekind's goal, for example, was to replace geometric proofs of results such as the monotone convergence theorem with ones that appealed only to a "purely arithmetic and perfectly rigorous foundation for the principles of infinitesimal analysis" (Dedekind 1901, pp. 1–2).

Axiomatic approaches to mathematics were in some sense a natural endpoint of the movement towards a rigorous methodology, with the only permitted inferences being logical deductions from a collection of basic principles or axioms given in advance. Geometry, with its Euclidean heritage, was a natural home for such an approach. Moritz Pasch in his 1882 lectures *Vorlesungen über Neuere Geometrie* (*Lectures on Modern Geometry*) (Pasch 1882) developed an axiomatic and deductive approach to geometry that was, unlike that of Euclid, “independent of figures”. Pasch’s work was a major influence on Hilbert (Schlimm 2010, p. 93), whose *Grundlagen der Geometrie* aimed to finish what Pasch had started and put Euclidean geometry on a fully axiomatic footing.[4]

However, to see Klein and Hilbert as members of two opposing camps, one championing intuition and the other championing rigour, does a disservice to both their works and their philosophical and foundational views. As Rowe (1989, 1994) argues, Klein had an appreciation for rigour and for the role that axiomatic thinking could play in mathematics, and was an early advocate of Hilbert’s *Grundlagen*. Hilbert, on the other hand, saw the need for intuition if we were to understand the axiomatic proofs; it can be argued that on Hilbert’s view, intuitive knowledge of Euclidean geometry was the chief motivation for pursuing an axiomatisation of geometry (see also Sinaçeur 1993, p. 260; Corry 1999, p. 157 and Grattan-Guinness 2000, p. 209).

Much like Klein and Hilbert, Afanassjewa clearly recognised the importance of both intuition and logical thinking in geometry: without intuition, geometrical thought is impossible, but without logical thinking, we are unable to properly regiment our intuitive thoughts and strip away inessential or misleading aspects. In her 1924 manifesto, Afanassjewa develops a foundational view of spatial intuition and geometrical reasoning—that is, of the epistemology of geometry—together with a theoretical and practical approach to teaching. Moreover, her approach to teaching is based not merely on her foundational account of geometry, but on a clearly developed philosophical picture of how human beings attain understanding in mathematics. It thereby bridges the gap between theoretical or abstract knowledge of geometry, and knowledge of the physical world. The exercises proposed in the *Übungensammlung*, and more generally her conception of the propaedeutic or introductory geometry course, are carefully designed in accordance with her more general views about how we come to learn geometry.

A distinctive aspect of Afanassjewa’s conception of intuition is that spatial intuition is not merely a given, but is instead a faculty which can be developed through systematic training. Unlike most other accounts of the faculty of intuition, both earlier and subsequent, Afanassjewa’s account does not proceed in analogy with perception, nor is it conceived of as a uniquely psychological or heuristic aid to the pursuit of mathematical knowledge. On her view, intuition is necessary for mathematical understanding, for problem-solving, and for rigorous thinking, but not sufficient for any one of these processes. In particular, Afanassjewa takes understanding to be a fundamental aspect of knowledge, and in this sense, intuition is also necessary for geometrical knowledge. While intuition can provide imprecise or inconsistent pictures, knowledge of abstract statements is ultimately achieved through elaborating the pictures provided by intuition by logical thinking, by isolating the salient aspects of the intuitive pictures through abstraction and ordering them logically to obtain a better—more precise, general, and consistent—intuitive picture than the initial one.

[4]More recent research in formalised geometry suggests that Hilbert’s axiomatic treatment in the *Grundlagen* still contained gaps that had to be filled by diagrammatic reasoning and geometrical intuition: see Meikle and Fleuriot (2003).

Several important themes emerge from Afanassjewa's proposals for reforming the teaching of geometry. One is her deep concern for foundational issues: What is the role of intuition in the teaching and learning of geometry? What is the role of logic? How do intuition and logic relate to understanding? Should we conceive of geometry as the science of space, or as an axiomatic system, or both? And if both, how do these two aspects interact? Another is her desire for an account of geometry as a practice carried out by both researchers and students, as opposed to a collection of theorems of a particular area of mathematics: this is evident in the key role played in her view by rigour and clarity in thinking and writing on the one hand, and understanding and insight on the other.

Afanassjewa's ideas are striking in their range, and in their richness. Despite being expressed in a concise way, because she was engaging with educators and not philosophers, her conception of geometry is clearly the product of a deep philosophical analysis and informs all of her writings on the subject. Nonetheless, there has been no substantial discussion from a philosophical perspective of Afanassjewa's views of the roles of intuition and logic in geometry. The first English translation of her 1924 manifesto provides an ideal opportunity to begin such a discussion. This paper examines the ideas contained in her manifesto, also taking in account her dispute with Dijksterhuis in the pages of the journal *Bijvoegsel van het Nieuw Tijdschrift voor Wiskunde* that immediately followed the publication of the manifesto, as well as her 1931 collection of exercises for an introductory geometry course, the *Übungensammlung*. Its aim is to provide a clear presentation of Afanassjewa's ideas about intuition, understanding, and geometry education, from a philosophical point of view. In doing so, it aims to help make the philosophical aspects of her work better known, stimulate a growing interest in her ideas, and act as a starting point for further research. In particular, it would be valuable to develop a more encompassing analysis of her work by taking into account how she influenced, and was influenced by, contemporary work on mathematical education in Russia, and how her views on mathematical education developed after the manifesto and the *Übungensammlung* in her later work in Dutch, including through her interactions with Freudenthal (for a discussion of her influence on Freudenthal see La Bastide-van Gemert 2015; Smid 2016; for a discussion of her legacy in mathematics education in the Netherlands see De Moor 1993; Smid 2009; Furinghetti et al. 2013).

Afanassjewa was a strong advocate of the cultural importance of teaching geometry in schools as a way to develop and train rigorous reasoning skills. The next sections will introduce her view, starting from the initial question of what makes geometry education valuable for everyone, regardless of their future educational and career trajectory.

2.3 The Value of Geometry Education

Both the manifesto and the *Übungensammlung* open with a compelling defence of the value of geometry education in schools. According to Afanassjewa, the value of teaching a given school subject resides in the transferability of the methods of treating problems in that subject: when a student internalises the methodology of that subject, they should be able to use it fruitfully in other areas of reasoning. The reason why it is desirable to teach geometry in schools thus depends on the reason why geometrical methods can be fruitfully applied in other areas of reasoning. This is best explained by distinguishing between two main ways in which mastering geometrical method is useful to everyone, including those who will not proceed to further study or an occupation that involves geometry or mathematics in any way.

First, the ability to adequately see and visualise space—i.e. to be familiar with spatial relations—is useful to everyone, because it is valuable to perceive and manipulate spatial relations as fast and accurately as possible in the many different situations in which we encounter spatial problems. These range from instinctively making a quick movement to avoid a sudden danger, to moving furniture and figuring out whether an oddly shaped sofa can pass through a doorway; from performing creative activities such as drawing, sculpting, or designing and sewing clothes, to the aesthetic enjoyment of different forms of art and of architecture, or the study of school subjects such as geography, mechanics, physics, etc. The need to possess adequate knowledge of spatial relations arise from our interaction with and experience of the physical world, and therefore merits special study: spatial imagination should be developed from childhood just as much as a musical ear or physical skills (Ehrenfest-Afanassjewa 1924, pp. 1–2).

Second, the geometrical method is characterised both in historical and in contemporary mathematics by its *striving for the utmost clarity* (Ehrenfest-Afanassjewa 1931, p. 3). What different strategies for problem-solving in geometry have in common is that a given problem is not set aside until it has become entirely transparent, the result is formulated in the clearest manner, and all the results thus obtained form a coherent, well laid out system. Since harmony, coherence, and logical order are the methodological ideals of any intellectual enquiry, it is clear that mastering the geometrical method has a very high educational value (Ehrenfest-Afanassjewa 1931, pp. 2–3). However, one might object that the former aspect of the value of geometry education is independent of the latter: knowledge of geometrical theorems is not necessary for knowledge of spatial relations, and indeed, the opposite may be true. When using spatial visualisation in action, knowledge and thinking often delay action or lead to less effective action: as Afanassjewa acknowledges, Pythagoras' theorem is of little use to "a painter, a cox, a hunter, or a cyclist (Ehrenfest-Afanassjewa 1924, pp. 1–2)".[5] Even in contexts in which the knowledge of certain quantitative spatial relations is important, it could be argued that the Euclidean framework is disposable, and that the development of spatial visualisation and the practice of its applications would be more useful than the teaching of Euclidean geometry. These considerations show that the question of whether Euclidean geometry should be taught in schools, and if

[5] It should be noted that on Afanassjewa's view, not only is knowledge of geometrical theorems not necessary for knowledge of spatial relations, it is also not sufficient. Her objections to the teaching of geometry by means of an axiomatic presentation of Euclidean geometry, and the relation of such a course to the development of spatial imagination, will be discussed below.

so, whether it should be presented axiomatically or through practical applications, is independent of the question of how best to develop spatial imagination.

The reason why Euclidean geometry should be taught in schools has rather to do with the most distinctive aspect of the geometrical method, namely, that it is characterised both in historical and in contemporary mathematics by its striving for the utmost clarity. The axiomatic treatment of spatial relations in geometry has reached a particularly high level of logical rigour, and it is thus to be expected that familiarity with geometry should have a positive effect on students' reasoning abilities, thereby demonstrating the particular cultural value of training in geometry (Ehrenfest-Afanassjewa 1924, p. 2). In this, Afanassjewa agrees with E. J. Dijksterhuis, the most vocal opponent of her manifesto. Dijksterhuis stresses the role that learning mathematics plays in learning to think, which he calls "the most precious fruit of a mathematical education [. . .] the purity and honesty of mathematical thinking and speaking [. . . and] the spiritual discipline, order and purity that mathematics pursues" (Dijksterhuis 1924a, p. 11, author's translation). Dijksterhuis's strong negative reaction to Afanassjewa's work should therefore not mislead us into thinking that their disagreement also extended to the ultimate value of geometrical education (La Bastide-van Gemert 2015, p. 139).

According to Afanassjewa, it is no coincidence that spatial relations have been the focus of deep logical analysis: we attach a high value to the organisation of our own thoughts in a logical manner, obtained by reflecting on our experience, and we all at some point experience the desire to express our experience and communicate it effectively to others (Ehrenfest-Afanassjewa 1924, p. 3). Doing so involves expressing clearly what we see intuitively, in a way that brings out what is essential to the speaker. This ability is what makes and has made scientific practice, including scientific collaborations and the sharing of scientific knowledge, possible throughout the centuries. In this respect, geometry enjoys the simplest and clearest form among the subjects of human thought, perhaps with the exception of the rest of mathematics. With respect to mathematics, however, the material of geometry has the unique benefit of being given in perception, however imperfectly, and is thus familiar to every human being from their own experience of the physical world.

It is this cultural value that justifies the teaching of Euclidean geometry as part of the school curriculum, including teaching it to people who do not have an aptitude for mathematics and who will not come into contact with mathematics or its applications in their future life. The fact that the value of mathematics education is doubted by many is, for Afanassjewa, to be ascribed to the current curriculum and exam requirements, which lack precisely those features that would allow the study of mathematics in schools to play its proper role in the development of students' thought. What these features consist in is the main subject of the manifesto and the *Übungensammlung zu enier geometrischen Propädeuse*. In the next section, I will present Afanassjewa's conception of scientific practice and geometric methodology, with a focus on the role that intuition plays in these contexts.

2.4 Intuition, Understanding, and Logical Thinking

Although many mathematicians in the nineteenth century, particularly within analysis, worked to develop a new standard of mathematical rigour that eliminated appeals to geometrical intuition, an opposing school of thought held fast to the value of intuitive reasoning (Torretti 1978; Gray 2010, 2019). Figures such as Klein, Poincaré, and Brouwer occupied a prominent position among those promoting a new, more positive attitude towards mathematical intuition (Glas 2002; Gray 2008; Heinzmann and Stump 2017; Gray 2019). Tatiana Afanassjewa's view sits squarely within this current of thought.

2.4.1 The Role of Intuition

According to Afanassjewa, "*Without intuition no thinking is possible*" (Ehrenfest-Afanassjewa (1924), p. 4, emphasis in the original). Intuition is what provides the mental representation of a certain material, which is then processed by conscious thought. More precisely, she argues that two steps should always be distinguished when acquiring insight into a certain issue: (1) seeing a certain feature in the intuitive picture that we have in mind, and (2) bringing this feature to awareness. The latter—which Afanassjewa calls *logical thinking*—is fundamental to all the steps of the reasoning process such as grasping and ordering what is initially represented in our intuitive picture, identifying the gaps and contradictions in it, attempting to fill those gaps, and finding the origin of any inconsistencies. The former, i.e. the material that is processed in the way just described, is the *intuition*. The content of intuition can be *concrete*, i.e. consisting of sensory perceptions, or *abstract*, i.e. consisting in the result of a previously analysed representation. Intuition can, if not processed logically, lead to mistakes and inconsistencies; this is not a fault of intuition itself, but rather a failure to employ appropriate reasoning tools to treat the intuitive material logically. Not every intuitive element of our representations has a place in "the *complete* intuitive picture" (Ehrenfest-Afanassjewa (1924), p. 4, emphasis in the original), and elements that are in contradiction with the whole should be identified and replaced by an appropriate, yet intuitive, element. It is compatible with this conception of intuition that this intuitive material can be communicated without the awareness proper of the process of logical thinking, and also that it can be ordered, albeit in an unconscious way. Afanassjewa stresses that this experience is commonplace in mathematical practice, and she recalls Gauss's famous saying that theorems were clear in his mind a long time before he knew how to find a proof, and an observation by Poincaré about the necessary role of intuition in searching and finding mathematical facts before a proof is reached (Ehrenfest-Afanassjewa 1924, p. 4; see also Footnote 3). This can happen because it is possible for someone to have a clear intuition of a certain subject and make purely logical, non-contradictory statements about the subject in question, and yet not be thinking logically in the sense discussed above. In such a situation, the failure in logical reasoning will be manifest once one can only acquire insight by searching it consciously: it will be impossible to reach and formulate clearly such insight unless an adequate logical analysis of the intuitive picture has already been carried out.

This view is in some respects close to Klein's view of the role of intuition. Klein also held that spatial intuition is central to both the learning and the teaching of geometry, and that it can be developed by experience, as becomes clear when one considers Klein's use of physical models of geometrical objects and structures as a prompt to develop spatial intuition.[6] However, on Klein's view any intuitive picture is essentially indeterminate and thus does not meet the standards of rigour of geometrical thought. In particular, intuition is not, by itself, sufficient to deliver precise geometrical notions or precise geometrical propositions: it can sometimes lead us to contradictory geometrical statements without providing sufficient means to decide between them. These issues can only be adjudicated by developing what Klein calls "refined intuition", i.e. by unfolding the logical consequences of exact axioms (see Klein 1893, p. 226; Mattheis 2019, p. 97 and Torretti 1978, Sect. 2.3.10, p. 147).

In a different way, Afanassjewa's view is also somewhat close in spirit to that of Poincaré, who thought that intuition (interpreted as an element of understanding) is necessary for mathematics both in the context of discovery *and* in the context of justification (Heinzmann and Stump 2017). Even though Afanassjewa does not distinguish explicitly between these two contexts and roles of intuition, both are present in her view: the role of providing a visual representation to be analysed and ordered by logical reasoning, and the role of providing insight in the process of solving a mathematical problem (for a discussion of the distinction between these two kinds of intuition and their role in different historical periods of mathematics, see Arana 2016). Note, however, that Poincaré had a different stance on geometrical intuition, maintaining that appeals to intuition should be completely dispensed with in geometry (for a discussion of Poincaré on geometrical intuition, see Heinzmann and Stump 2017 and Heinzmann 2013, as well as Gray (2008, 2019) for more general discussions on the epistemology of geometry).

Hence, Afanassjewa argues, both everyday experience and mathematical practice show that we can have insight—an intuitive picture of a given object—and even perform actions that are not guided by explicit logical reasoning, but that the opposite is not the case: logical thinking can *only* happen if an intuitive picture is provided first. Even when intuition imposes a certain degree of order on the intuitive representation, it is only by employing logical reasoning that such intuitive representations are clearly formulated and logically ordered (Ehrenfest-Afanassjewa 1924, p. 5). As seen above, an intuitive picture that has not been processed through logical reasoning is liable to contain mistakes and inconsistencies, and the constitutive aspect of logical thinking consists precisely in the bringing to awareness of the logical relations that hold between the statements that describe the intuitive picture. Without such material being provided by intuition, no logical thinking would be possible at all.

---

[6] For an analysis of the role of intuition in Klein's conception of geometry education, see Mattheis (2019) and Rowe (1985).

It may be argued, against Afanassjewa, that the process of obtaining scientific knowledge is not simply the process of logically treating our intuitive representations, but is in fact the process of *replacing* intuition by logic and insight by scientific explanation or mathematical proof (Ehrenfest-Afanassjewa 1924, pp. 4–5). Some of Afanassjewa's own observations seem to conform to this view. She explicitly claims that "The actual work of logic happens at those moments in which intuition is brought to consciousness": the main difficulty in formulating a mathematical proof is that of making the essential premises explicit, and once this is done, a clear presentation of the proof does not require comparable effort. A clearly formulated mathematical proof is the evidence that a given intuitive material has been adequately analysed: as such, mathematical proof is the mark of scientific reasoning in mathematics. In what sense, then, is intuition essential to science in general, and to mathematics and geometry in particular? Afanassjewa's answer to this question hinges on the role that she assigns to understanding within the attainment of knowledge.

2.4.2 Understanding and Logical Thinking

Mathematical proof, conceived of as a logical concatenation of inferences, is not, as Afanassjewa puts it, "the *instrument* itself of thinking (Ehrenfest-Afanassjewa 1924, p. 4)": neither problem-solving, nor mathematical understanding, stem from constructing or surveying a logical concatenation of inferences.

A proof constitutes the end result of a thinking process; sometimes, it can also constitute the beginning step for new research. However, looking for the solution to a mathematical problem does not proceed by concatenating inferences logically. According to Afanassjewa, problem-solving is the result of shaping a mathematical question in the most effective way, and becoming aware of what precisely is being searched for (Ehrenfest-Afanassjewa 1924, p. 5). As she also remarks in the *Übungensammlung*, nothing essentially new can be found by means of a prescribed recipe, precisely because the essentially new is unknown: we do not know where to find it, how to find it, or even that what we are looking for really exists (Ehrenfest-Afanassjewa 1931, pp. 2–3). For example, mastering Euclidean geometry does not give us any recipe for finding new directions of research in geometry. On this point, Afanassjewa agrees with Poincaré, who argues that "Pure logic could never lead us to anything but tautologies; it could create nothing new; not from it alone can any science issue" (Poincaré 1958, p. 19).

Analogously, mathematical understanding is not achieved merely by checking a rigorous proof step by step. For example, it is easy to see that there must exist a relation between two sides and an angle of a triangle, and its third side. On the other hand, it is not equally easy to see that in the case of a right angle this relation is expressed by the Pythagorean theorem (Ehrenfest-Afanassjewa 1924, Footnote 6). What produces understanding, instead, is seeing the intuitive connections at every step of the proof: merely checking a rigorous proof step by step without an intuitive understanding of each step of the proof will lead one to consider the statement in question temporarily proved, but not understood (Ehrenfest-Afanassjewa 1924, p. 4-5; footnotes 4 and 9). In brief, there can be no mathematical discovery and no understanding without intuition, and this is why intuition is necessary for science as much as systematic theories are.

What, then, is the relation between intuition and understanding, and logical thinking? Answering this question requires an analysis of the roles of two other reasoning processes that are necessary for mathematical proof, and ultimately, for scientific thought in general: *abstraction* and *calculation*. Intuition plays a role in both processes; the analysis of this role sheds light on the relation between understanding and the processes of abstracting and calculating.

A clearly formulated mathematical proof, as discussed above, is a sign that an issue has been deeply analysed by means of logical thinking. The process of treating an intuitive representation logically takes place when an intuitive picture is brought to awareness, namely, when one *abstracts* from the object intuitively represented by ordering the intuitive material and selecting only the relevant features, and when one considers the formal relations between the statements that describe the object (Ehrenfest-Afanassjewa 1924, pp. 5–6). More specifically, we abstract when we are interested in understanding a specific property of an object that we do not yet understand but that we think is important to understand, so we identify and isolate only those elements of the intuitive picture that are relevant to that property and its understanding, and put the other elements in the background. It is for this reason that intuition is also necessary for abstraction.

The subsequent operation of considering the formal relations between statements that describe the object is best characterised as *calculating*, rather than logical thinking (Ehrenfest-Afanassjewa 1924, p. 5). What Afanassjewa calls "calculating" is a key component of scientific research, and in this context, intuition is important in a number of ways. Firstly, the thinking relation between the agent and the object that is represented is an intuitive one. Secondly, the relations of subsumption between the statements in question and the formal–logical structure of each statement are themselves understood intuitively before they can be formulated consciously. Lastly, the gaps in the intuitive picture can only be closed by using spatial imagination. To illustrate the difference between logical thinking and calculating, Afanassjewa mentions the algebraic treatment of formulas in physics: thinking logically about the algebraic relations that hold between the formulas in question is not an instance of thinking logically about the physical relations themselves, but an instance of calculating algebraically certain formal relations that are represented in the formula. Hence, calculating cannot, by itself, lead to understanding.

While nothing essentially new is discovered by means of a proof, new consequences can be drawn by mere calculation or by manipulations according to the rules of formal logic. Knowledge obtained by mere calculation, however, is different from knowledge obtained by logical thinking because the latter always leads to certain conclusions, while the former can easily be wrong, unless, as in the case of intuitive representations, it is treated logically and assigned a place in the original intuitive picture, i.e. made consistent with it. From this we can see clearly the sense in which for Afanassjewa, logical thinking is distinct from the mere drawing of inferences by means of applying the rules of formal logic, and is a much more complex cognitive and epistemic process.

These considerations support Afanassjewa's view that intuition is necessary for science (and for mathematics in particular) not merely as the faculty of thought that provides the raw material to be processed by means of logical thinking, but also because intuition is necessary for understanding, which is necessary to the pursuit of knowledge. In the next section, Afanassjewa's conception of intuition and of logical thinking will be connected to her conception of geometry teaching, which reflects her view on the value of geometry education presented in Sect. 2.3.

2.5 The Study of Space and the Axiomatics of Geometry

Afanassjewa's distinction between intuition and logical thinking is reflected in her distinction between the study of space and the axiomatics of geometry, two disciplines that she argues have been conflated since Euclid's time. The intuitive material in the study of space is given by spatial imagination, whose content is constituted by sensory perceptions, while the intuitive material in axiomatics is given by the axioms, which are distilled from the spatial relations that are the subject matter of the study of space, or by the theorems, that result from a former logical elaboration (Ehrenfest-Afanassjewa 1924, pp. 5–6). But how do we get from sensory perceptions to an axiomatic system for geometry?

On Afanassjewa's view, building an axiomatic system means "creating order out of chaos" (Ehrenfest-Afanassjewa 1931, p. 4). This can be achieved by bringing together—not necessarily in a systematic way—the following elements: (a) isolation of the features that identify the area of research in question; (b) formulation of the initial problems with sufficient clarity; (c) identification of the primitive concepts, i.e. concepts to which the other concepts in the same area can be reduced; (d) identification of the elementary relations between these concepts—given in the axioms—from which all other relations can be deduced as logically necessary consequences; (e) identification of the most appropriate style of presentation for that system (Ehrenfest-Afanassjewa 1931, p. 4). In the case of geometry, we first logically analyse spatial imagination in order to identify the most fundamental spatial relations, and the spatial relations that go beyond our immediate spatial imagination. The truth of the statements that describe the most fundamental spatial relations is easily recognised on the basis of our experience of the physical world, while the truth of the statements that describe the spatial relations that go beyond our immediate spatial imagination is recognised by seeing the connections that hold between them and the fundamental spatial relations that are most familiar to us from experience. Secondly, these relations are then clearly formulated, so that only a few elements of the intuitive picture are "displayed", and their fundamentality is proved by showing that all the other spatial relations can be logically deduced from the fundamental ones. The result of this analysis is a logically coherent and elegant system of axioms.

For Afanassjewa, the statements that are eventually called "axioms" are *independent* statements, i.e. statements that cannot be logically derived from other axioms (Ehrenfest-Afanassjewa 1924, p. 6)—a conception that had been established by Hilbert's seminal work on the foundations of geometry (Hilbert 1899). Their function is to lay out the elementary relations between the fundamental concepts of an area of mathematics, thereby providing an insightful picture of that area that allows scientific communication. However, by carrying out the process of logical analysis of our sensory perceptions, our knowledge of space becomes both richer and clearer—much more so than the intuitive picture that we can form on the basis of the axioms, without having formulated them ourselves. In other words, "seeing the truth of the axioms is not the same as having them at one's disposal (Ehrenfest-Afanassjewa 1931, p. 4)": the Euclidean axioms will seem obvious to nearly everyone they are presented to, but this does not imply that those who recognise this are also aware of spatial laws and concepts. For example, it will not occur to the person who is unaware of spatial laws and concepts that the relevant spatial relations encountered in activities such as packing, moving furniture, cutting cloth, etc. can be reduced to planar problems. Hence, axioms are at one's disposal only when a deep understanding of the fundamental spatial relations has been achieved, and such relations have been logically ordered in the sense described above (Ehrenfest-Afanassjewa 1931, p. 6).

Despite the praise reserved for geometry and the geometric method, if the term "logical" is taken to indicate what is free from contradiction, well arranged, and transparent, Afanassjewa claims that it should only be used in connection with the Euclidean *style of presentation*, which orders the visually given material in such a way that the theorems follow one another naturally (Ehrenfest-Afanassjewa 1931, p. 2). As she writes in her manifesto (Ehrenfest-Afanassjewa 1924, p. 12), until the beginning of the twentieth century much emphasis was put on the formal–logical side of thinking, which was also emphasised in geometry education as shown by the school curriculum in several European countries including the Netherlands, where she had moved in 1912. The failure of this model of teaching geometry brought about a movement of teachers questioning the systematic approach to geometry education and aiming to teach geometry in a way that develops "intuition". However, both the systematic and the experimental sides of this debate are, according to Afanassjewa, guilty of the same sin: they focus only on the importance of one aspect of the faculty of thinking, either logical or intuitive, when each is just as necessary as the other to mastering geometrical thinking. This misconception arises from their shared and overly narrow conception of the subject matter of geometry and the according value of geometry education, namely, to teach students the contents of certain theorems of Euclidean geometry (Ehrenfest-Afanassjewa 1931, pp. 1–2).

Instead, according to Afanassjewa, the aim of geometry education is to achieve an intuitive picture of the subject that is much more developed and precise than the initial one, and that is well understood (Ehrenfest-Afanassjewa 1924, p. 6). Thus, she argues that geometry education can be most useful to the development of both spatial imagination and logical thinking only when intuition is accorded its proper role in the process of thinking (see e.g. Ehrenfest-Afanassjewa (1924), pp. 2–3).

2.5.1 The Systematic Approach

The systematic approach to geometry education advocates teaching geometry axiomatically, i.e. as it is presented by Euclid in the *Elements*. The usual assumptions behind this approach are either that understanding of spatial relations is obtained synthetically with the help of the axioms of geometry, or that the necessary spatial intuitions are possessed innately by everyone, and it is thus sufficient to build on that by unfolding the proofs before the students. As discussed earlier in this section, Afanassjewa believes that the value of axiomatics consists in revealing the logical relations between geometrical propositions, and that for this reason the Euclidean style is the most appropriate way of presenting geometry, and a paradigm of logical thinking. This motivates Afanassjewa's belief that axiomatics is a key component of the study of geometry, and her belief in the high practical value of making geometry part of the school curriculum (see also Sect. 2.3 above). Since, however, the axiomatic method is not the method by which understanding is achieved, teaching Euclidean geometry axiomatically will not lead to the desired results of teaching students fundamental facts about spatial relations and providing them with the ability to think logically (Ehrenfest-Afanassjewa 1924, pp. 8–9), contrary to what the advocates of the systematic approach to geometry education, such as Dijksterhuis, argued in response to the manifesto (Dijksterhuis 1924b; De Moor 1993; de Moor 1996). According to Afanassjewa, the desired results can only be obtained if the geometry course is designed not only to expose students to logical thinking in the manner of Euclid, but also to create the basis for logical thinking, namely, a sufficiently trained spatial intuition and the curiosity to analyse it logically.

Afanassjewa identifies three factors, whose roots lie in the current curriculum, that contribute to students' lack of interest in logical thinking and the axiomatics of geometry. The first is new or misleading terminology: for example, the term "proof" is used in the study of space to indicate an argument that provides insight into the correctness of a proposition whose truth is already accepted on the basis of our sensory experience. In the axiomatics of geometry, on the other hand, the term "proof" indicates an argument that logically traces the theorem back to the axioms. This means that when taking the systematic course, the students may find themselves in the position of accepting that a certain theorem has been proved and yet not realise that it is valid of the physical world. Alternatively, they may not appreciate why a completely obvious statement about physical space requires proof. This is sometimes obviated by the teacher by encouraging the students to doubt their knowledge of that statement in order to see it eventually established by rigorous proof (Ehrenfest-Afanassjewa 1924, p. 7; Ehrenfest-Afanassjewa 1931, p. 3). However, according to Afanassjewa, this method only produces the effect of making the students doubt science and the scientific method in general, regardless of the fact that different students have different ideas about what statements count as obvious. Such obstacles can be overcome by making the students acquainted with the statements that are axiomatised in Euclid's system prior to their study of axiomatic geometry, i.e. by teaching the students the study of space before they get to study the axiomatics of geometry (Ehrenfest-Afanassjewa 1924, pp. 7–8).

The second factor discouraging students from being interested in the axiomatics of geometry is the lack of spatial imagination. It is not surprising that axiomatic proofs are not learnt by the students when proofs are taught, e.g. by making students repeat them and waiting until they become accustomed to the proving style, and thereby become clearer thinkers. The reasons are those discussed in Sect. 2.4.2: Euclidean geometry presents an exceptionally clear summary of the results of logical thinking, but it does not reproduce the process of thinking itself. In order for the students to benefit from learning the axiomatics of geometry, the students must first achieve the results by themselves (Ehrenfest-Afanassjewa 1924, pp. 8–9).

The third factor concerns the logical elaboration of the material and the risk of overloading the course. It is better to cover less material but in a more careful and detailed way, i.e. in one that makes the student appreciate the benefit of axiomatics and teaches them logical thinking that can then be applied in other areas. If, on the contrary, the student learns several theorems but does not understand them, then when confronted with a new problem the student will use a formula that has been successful before, as opposed to working through the problem themselves. This problem can be overcome by explaining the goals of axiomatics at the start of the course and by making students acquainted with spatial relations before studying the axiomatics of geometry, so that the systematic course presents itself as a useful and elegant analysis of solutions to practical problems concerning physical space (Ehrenfest-Afanassjewa 1924, pp. 9–10).

Does the experimental approach address these issues adequately?

2.5.2 The Experimental Approach

The experimental approach (or "laboratory method of education") aims to develop spatial imagination by teaching geometrical propositions empirically, i.e. by making students practice the application of geometrical theorems to practical problems (for example, through measuring, cutting, and drawing), thereby verifying their correctness (Ehrenfest-Afanassjewa 1924, section 8.4, and Ehrenfest-Afanassjewa 1931, section 5). While Afanassjewa praises this approach as coming closer to developing students' spatial intuition than the systematic approach, in her view, it suffers from serious limitations.

A major limitation of this approach is that it fails to distinguish the contribution that the visual investigation of a theorem brings to the development of spatial imagination from the contribution that the manipulation of measurement brings to the verification of the theorem. Very often, measurement does not contribute to the development of spatial imagination: e.g. when students are asked to measure a circumference with a tape measure and determine how many times the diameter is contained in it, the visual picture does not provide any compelling reason for the ratio to be one number rather than another. On the other hand, when the students are asked to inscribe a regular hexagon in a circle, they can thereby both *see* and *prove* that the hexagon's perimeter is equal to 6 radii, but smaller than the circumference

(Ehrenfest-Afanassjewa 1931, pp. 8–9, fn. 1). That is, in order to internalise a visual proof completely, the student needs to already possess a sufficiently developed spatial imagination. If this is missing, the student will carry out the proof mechanically, and since the proof is then immediately followed by an application that is usually computational in character, rather than visual, the proof learnt after the manipulation of measurement will not contribute to the development of spatial imagination (Ehrenfest-Afanassjewa 1931, section 5). So, Afanassjewa argues, if a student comes to the truth of a theorem heuristically, the joy of learning that derives from the mental effort of reaching the solution to a problem will be missing, and as a result the student will not come to the systematic exposition of the relevant result with curiosity and interest (Ehrenfest-Afanassjewa 1924, section 8.3–8.4; Ehrenfest-Afanassjewa 1931, pp. 8–9).

Another limitation of this approach is that it cannot be expected that students will draw the correct generalisations from their experience in applying geometrical theorems. Returning to the exercise just mentioned above, it is likely that the students' measurements of the circumference with a tape will yield different results at different times, and the teacher's statement to the class of the right result at the end of the exercise will not produce an understanding of the general theorem in the individual student (Ehrenfest-Afanassjewa 1931, pp. 7–8). That is, sometimes measurement does not even adequately contribute to the verification of the theorem in question by the students. Furthermore, if students have only acquired familiarity with individual theorems through practising their application, when attempting to solve a problem they are likely to recall from their memory a specific formula that was useful in solving a specific problem, as opposed to working through the problem themselves. As seen above (Sect. 2.5.1), this problem also arises in the context of the systematic approach, and for the same reason, namely, the focus on teaching the content of particular theorems rather than developing spatial intuition. A predictable consequence of this phenomenon is that students lose the thread when they come to learn more complex theorems, and that only a few disconnected theorems will be retained in their minds after the end of their studies (Ehrenfest-Afanassjewa 1924, p. 9).

Empirical work such as measuring, drawing, sculpting, cutting, and pasting has an important place in Afanassjewa's plan for geometry education, but it must be prompted by the development of spatial imagination rather than by the verification of theorems, and accompanied by the logical analysis of spatial intuitions. In other words, the aim of geometry education should not be teaching the correctness of given geometrical theorems, but providing the students with the fundamental geometrical concepts and the ability to mentally manipulate them. If prompted with the right exercises, the subconscious mind retains impressions that are necessary to later recognise the truth of geometric theorems.

For this reason, the development of the student's own thinking process and imaginative skills should be prioritised over the reproduction of knowledge of individual theorems. Students should first try to imagine the geometric figure under consideration, and only afterwards the teacher should test and correct the students' imagination to match the real objects (Ehrenfest-Afanassjewa 1924, p. 13; see also De Moor 1993). In this sense, the experimental method should only be used with the aim of getting the students to start to guess a certain regularity in the existence of spatial relations, laying the ground for them to want to establish those more rigorously later on.

2.5.3 The Propaedeutic Course and the Systematic Courses

To overcome the shortcomings of both the systematic and the experimental approaches, Afanassjewa suggests that the students first take a *propaedeutic* or *introductory course*—the study of space—whose aim is to develop spatial imagination (see chiefly Ehrenfest-Afanassjewa 1924, section 8.4). The course is designed to provide the students with a repertoire of geometric images and the ability to mentally manipulate them, so that the fundamental geometrical concepts can develop from concrete experience of the physical world, and the geometric terminology can be connected to the mental representations of the geometric images. Hence, the students should not prove any theorems at this stage, when they do not yet have the appropriate instruments to deeply understand them, but should instead focus on doing exercises with the goal of learning key geometrical concepts (Ehrenfest-Afanassjewa 1931, p. 5). In this context, students should not be taught terminology and diagrams in a way that is disconnected from their familiar experiences. Geometry should be presented as a discipline that operates with concepts that can be obtained by abstracting from one's sensory experience; terminology should be associated with the key geometrical concepts that are developed from spatial intuition, and diagrams should be used to generalise spatial relations observed in experience, rather than constituting a self-standing object of study for the students (Ehrenfest-Afanassjewa 1931, pp. 9–10).

The intuitive content for the study of space is given by sensory experiences: Afanassjewa stresses that regardless of the educator's view on the aprioricity of spatial representations, the students will find it easier to visualise in their mind those spatial relations that are already familiar to them from experience than if they are presented with Euclidean proofs from the beginning. A vivid interest in geometry will arise in many students once they recognise how densely spatial problems penetrate everything we do or know (Ehrenfest-Afanassjewa 1931, p. 6). If students are receptive to appreciating the possibility of fitting individual theorems into a simple system whose axioms convey the fundamental truths about spatial relations, then they will develop an interest in a more systematic approach to geometry by the time they come to study geometry axiomatically. As seen in Sect. 2.5, this can only be obtained if students' spatial imagination is already suitably trained in manipulating geometrical objects mentally and they have suitable familiarity with the fundamental facts concerning spatial relations. Therefore, while visual support material is of crucial importance at the beginning of the course, the teacher should not rush to give the students visual support material as they progress in the course, but should instead encourage them to use their own spatial imagination (Ehrenfest-Afanassjewa 1931, p. 12). Since the propaedeutic course constitutes a key step in the learning of geometry and the development of spatial imagination, time should not be saved at this stage for the necessary mental manipulations: the laboratory exercises should therefore last as long as needed, based on the level of development of the spatial imagination of the class, and no longer than needed in order to prepare the students for the systematic course.

Subsequently, the students should take a *systematic course*, whose main aim is to develop logical thinking. This course differs from the traditional ones in that propositions that are evident to all students in the class should not be proved but assumed temporarily as axioms. Other propositions will be formulated and proved under the guidance of the teacher, but with a substantial input from the students (Ehrenfest-Afanassjewa 1924, pp. 10–11). This course is designed to make the student come to appreciate the value of rigorous proof, where such appreciation is unlikely to be triggered by going through proofs of statements that the students already find obvious. The aim of developing logical thinking is thus cultivated in the systematic course by making the students formulate clearly their own perceptions and order them logically, and it is facilitated by the intuitive presentation of the material that draws on what the students learnt in the propaedeutic course. It is important that this course is kept as concise as possible, in order to give as much space as possible to the development of spatial imagination and logical skills.

Finally, students should take a *revision course* that recaps what has been learnt in axiomatic geometry in the systematic course. At this stage, the obvious propositions that were not proved in the systematic course—the "temporary axioms"—are proved, and the logical dependencies among the propositions are analysed. The purpose of rigorous proofs at this stage is not to establish the truth of general theorems, whose truth is already recognised, but to establish the logical dependence of given theorems on other theorems: the chief aim of logical thinking is to concentrate the subject of investigation and connect the whole area using a small number of connections (Ehrenfest-Afanassjewa 1924, pp. 10–11). For this reason, determining which theorems are axioms (i.e. which proposition is independent of the others but allows the derivation of all the other theorems) should be left to the end of the geometry course, and only offered to the most talented students who are curious about these matters (Ehrenfest-Afanassjewa 1931, pp. 13–15).

The goal of the teacher is thus to bring the student from the realisation that space presents itself in a chaotic form, to familiarity with the fundamental spatial relations, to the appreciation of rigorous proofs that allow to fit a number of theorems into a straightforward and elegant system of axioms by showing the progress from chaos to axiomatic system which comes with the systematic treatment of the subject. In order for the teaching method to be maximally effective, it is necessary that the virtues of the geometric method in solving problems and systematising spatial relations are experienced personally by each student, and more importantly, that the mastery of this method is seen as desirable by each student (Ehrenfest-Afanassjewa 1924, pp. 12–13, and Ehrenfest-Afanassjewa 1931, p. 4).

To this end, Afanassjewa's manifesto includes 14 exercises, followed by 194 exercises in the *Übungensammlung*. The course outlined in the *Übungensammlung* is not structured linearly, but arranged in 19 subjects aimed at developing different concepts in the theory of space. The activities in both the *Manifesto* and the *Übungensammlung* are of different complexities, and their sequence is intended to be chosen by the teacher in a way that matches the level of development of spatial imagination of the students in the class. They include activities such as estimating distance, angle, length, and proportion, visualising straight lines as axes of rotations or rays of light, or flat shapes as arising from manipulations of three-dimensional objects,[7] imagining endless lines that continue through walls and buildings, making schematic drawings of simple mechanisms, determining what data is necessary in order to solve a given geometrical problem, comparing the results of exercises carried out on planar and spherical surfaces, respectively,[8] and many more. Other exercises involve notions such as symmetry, shadow, perspective, as well as topological notions.

In the concluding section, I will briefly discuss how these ideas influenced geometry education in the decades following the publication of the manifesto and the *Übungensammlung*.

2.6 The Reception of Afanassjewa's Approach to Teaching Geometry

Afanassjewa's manifesto met with an immediate and somewhat hostile reaction from the Dutch community of mathematics teachers and mathematicians interested in education. Particularly notable was the reaction of mathematics teacher and historian of mathematics E. J. Dijksterhuis. His response to the manifesto led to the founding of a supplement to the mathematics journal *Nieuw Tijdschrift voor Wiskunde*, *Bijvoegsel van het Nieuw Tijdschrift voor Wiskunde*, dedicated to mathematical education, and later renamed *Euclides*. Dijksterhuis's article (Dijksterhuis 1924a) was published in the first issue of the *Bijvoegsel*, followed in the second issue by a reply from Ehrenfest-Afanassjewa (1924b) and a further response from Dijksterhuis (1924b).

Dijksterhuis was a proponent of the systematic approach to geometry education discussed in Sect. 2.5.1. A major source of disagreement between him and Afanassjewa was thus the role of proof in geometry education: invoking Dedekind's principle that nothing capable of proof should be accepted without it, Dijksterhuis argued that geometry should be taught axiomatically in the style of Euclid's *Elements*. In so arguing, he rejected Afanassjewa's claims that while Euclidean axiomatics is the best way of systematising the subject matter of geometry, understanding does not proceed by surveying axiomatic proofs. By rejecting an essential role for intuition in scientific thinking, Dijksterhuis sided with the tradition of which Dedekind was one

[7] A sample exercise of this kind is the following: "What form does a surface of water have in a cylindrical glass held at different angles?" (Ehrenfest-Afanassjewa 1931, p. 32). Such an activity is clearly aimed at developing projective imagination. Even though Afanassjewa does not discuss it as explicitly as Klein, training this aspect of spatial imagination is nevertheless important in her view.

[8] Afanassjewa was fascinated by the discovery of non-Euclidean geometries, their respective axiom systems and associated models, which she saw as a possible subject of study in high school.

of the main exponents and which aimed to dispose of any appeal to geometric intuition (see Sect. 2.2). Dijksterhuis, however, was in fact advocating for a far stricter approach in mathematical education than Dedekind himself, who observed that while the use of geometrical intuition was not completely rigorous, it was "exceedingly useful, from the didactic standpoint, and indeed indispensable, if one does not wish to lose too much time" (Dedekind 1901, p. 1).

In the decade following the publication of her manifesto, Afanassjewa's ideas appear to have had little impact on the teaching of geometry in the Netherlands, which remained stuck in the proof-based Euclidean paradigm advocated by Dijksterhuis and the other "logicians". By the mid-1930s, however, more voices in the Netherlands were calling for reform, amongst whom were the members of the *Wiskunde Werkgroep* (Mathematics Working Group), established in 1936 under the auspices of a larger educational reform movement led by Kees Boeke.[9] The Wiskunde Werkgroep met at Afanassjewa's house in Leiden, effectively as a continuation of the seminars she had held there in the 1920s, after she moved there from Saint Petersburg, and until the Second World War. In 1947, the Werkgroep began to be attended by Hans Freudenthal, who had moved to Amsterdam in 1930 as an assistant of L.E.J. Brouwer and was then newly appointed to a chair in mathematics at Utrecht University in 1946. Freudenthal, who would go on to become one of the leading figures in mathematical education in the Netherlands, was strongly in favour of an intuitive approach to geometry, and praised Afanassjewa's work on the teaching of geometry, especially the exercises of the *Übungensammlung*, which he called a masterpiece (van Hiele 1975; De Moor 1993; Smid 2016).

In the years after the war, Afanassjewa was still one of the leading figures in the Werkgroep, and many of its meetings were still held in her home (La Bastide-van Gemert 2015, p. 117). Through these meetings, perhaps even more than through her publications, Afanassjewa brought to the Netherlands an international perspective on mathematics education, at a time when mathematics education was not yet a research field and the emerging debate was largely confined to national contexts. Unlike the situation in mathematical research, where international collaboration and dissemination was the norm, debates on different aspects of mathematical education were by their nature tied to national issues of curriculum and the structure of the school system in a given country, and moreover were conducted in the language of the country in question (Furinghetti et al. 2013; Smid 2009, 2012; Nabonnand 2007).

Afanassjewa's interaction with Freudenthal led to a joint production: the 1951 pamphlet *Kan het wiskundeonderwijs tot de opvoeding van het denkvermogen bijdragen?* (*Can mathematics education contribute to the education of the intellectual capacity?*). The structure of the pamphlet was in some ways similar to that of Afanassjewa's debate with Dijksterhuis in the *Bijvoegsel*: an initial article by Afanassjewa was followed by Freudenthal's response, and then by further responses from both parties. Freudenthal took issue with the main point of agreement between Afanassjewa and Dijksterhuis, namely, that mathematics was an ideal subject to aid the

[9] For a brief history of the Wiskunde Werkgroep, see La Bastide-van Gemert (2015, pp. 30–32). See also Smid (2009); Furinghetti et al. (2013).

improvement of students' intellectual capacity: in short, that learning geometry was a good way to learn how to think (De Moor 1993, p. 19, La Bastide-van Gemert 2015, p. 139). As discussed earlier in this paper, Afanassjewa held that learning mathematics was an ideal way in which to develop the intellectual capacity, or "good thinking". This included not merely the ability to make inferences on the basis of a given set of premises, but a broader range of cognitive powers including

> seeking the essentials in a given situation (power to abstract)[;] realizing this (which is not the same); trying to formulate this sharply; confronting an obvious answer to a question with the total range of data and never losing sight of the total situation (ability to criticize).[10]

Elementary geometry was particularly suited to this educational role because of its simplicity, and because our experience with physical space and spatial relations gives us a way to approach the subject on an intuitive basis. Freudenthal, however, objected that this very simplicity was in fact what made mathematics problematic as a subject promoting the development of critical thought, since its "overly simple structure protects it against the occurrence of inconsistencies" (Ehrenfest-Afanassjewa and Freudenthal 1951, p. 25). Freudenthal further argued against Afanassjewa's views on more practical grounds, writing that

> I do not deem it impossible that someone, practising a certain subject, also learns to think in a more general sense [. . .] I do not at all exclude the possibility that mathematics contributes to practicing thinking. But I fear that one is building on quicksand when one wants to justify the periods that some school subject requires by making an appeal to the thinking exercises that those periods would be devoted to.[11]

Nevertheless, in this exchange Freudenthal seems to a large extent to have been talking past Afanassjewa, and addressing the question of whether mathematics should be used to train students to think, rather than the question of whether it could (see van Hiele 1975 and La Bastide-van Gemert 2015, p. 139).

Despite this disagreement, Freudenthal was deeply influenced by and sympathetic to many of Afanassjewa's views (Smid 2016). De Moor (1993) goes so far as to argue that Afanassjewa was the driving force behind the development of Freudenthal's views on mathematical education, citing a number of substantial parallels in their ideas: the emphasis placed on the individual learning of the student, the appeal to spatial intuition as a basis for learning geometry, the connection with physical reality, and the development of deductive reasoning. Ultimately, in large part due to Freudenthal's influence, Afanassjewa's ideas entered into the primary and secondary mathematics curriculum in the Netherlands. De Moor (1993, p. 22) reports that "more than sixty years after the publication of the *Übungensammlung* [in the curriculum that came into force in 1992], there is now an intuitive introduction to geometry".

---

[10]Ehrenfest-Afanassjewa and Freudenthal (1951, p. 6). The English translation of this passage is drawn from La Bastide-van Gemert (2015, pp. 136–137).

[11]Ehrenfest-Afanassjewa and Freudenthal (1951, p. 16). The English translation of this passage is drawn from La Bastide-van Gemert (2015, p. 137).

References

Arana, A. (2016). Imagination in mathematics. In A. Kind (Ed.), *The Routledge handbook of philosophy of imagination* (pp. 463–477). Routledge.

Bussi, M. G. B., Taimina, D., & Isoda, M. (2010). Concrete models and dynamic instruments as early technology tools in classrooms at the dawn of ICMI: From Felix Klein to present applications in mathematics classrooms in different parts of the world. *ZDM*, *42*(1), 19–31. https://doi.org/10.1007/s11858-009-0220-6.

Boi, L., Flament, D., & Salanskis, J. M. (Eds.) (1992). *1830–1930: A century of geometry: Epistemology, history and mathematics*. Number 402 in lecture notes in physics. Berlin, Heidelberg: Springer. https://doi.org/10.1007/3-540-55408-4.

Born, M., & Born, H. (1969). In A. Hermann (Ed.), *Der Luxus des Gewissens: Erlebnisse und Einsichten im Atomzeitalter*. Nymphenburger Verlag, Munich.

Chislenko, E., & Tschinkel, Y. (2007). The Felix Klein Protocols. *Notices of the American Mathematical Society*, *54*(8), 960–970. https://www.ams.org/journals/notices/200708/200708FullIssue.pdf.

Corry, L., & Hilbert, D. (1999). Geometry and physics: 1900–1915. In J. J. Gray (Ed.), *The symbolic universe: Geometry and physics (1890–1930)* (pp. 145–188). Oxford: Oxford University Press.

De Moor, E. (1993). Het "gelijk" van Tatiana Ehrenfest-Afanassjewa. *Nieuwe Wiskrant*, *12*(4), 15–24.

de Moor, E. (1996). Het ongenoegen van Dijksterhuis. *Nieuwe Wiskrant*, *15*(4), 22–26.

Dedekind, R. (1872). *Stetigkeit und irrationale Zahlen*. Vieweg, English translation in Dedekind (1901).

Dedekind, R. (1888). *Was sind und was sollen die Zahlen?* Vieweg, English translation in Dedekind (1901).

Dedekind, R. (1901). In W. W. Beman (Ed.), *Essays on the theory of numbers*. Open Court, English translations of Dedekind (1872) and Dedekind (1888).

Dijksterhuis, E. J. (1924a). Moet het Meetkunde-onderwijs gewijzigd worden? Opmerkingen naar aanleiding van een brochure van Mevr. Ehrenfest-Afanassjewa. *Bijvoegsel van het Nieuw Tijdschrift voor Wiskunde*, *1*(1), 1–26. https://archief.vakbladeuclides.nl/bestanden/001_1924-25_01.pdf.

Dijksterhuis, E. J. (1924b). Antwoord aan mevrouw Ehrenfest-Afanassjewa. *Bijvoegsel van het Nieuw Tijdschrift voor Wiskunde*, *1*(2), 60–68. https://archief.vakbladeuclides.nl/bestanden/001_1924-25_02.pdf.

Ehrenfest-Afanassjewa, T. (1924a). *Wat kan en moet het Meetkunde-onderwijs aan een niet-wiskundige geven?* Paedagogiese Voordrachten. J.B. Wolters, Groningen and the Hague, English translation in this volume by P. A. van Wierst.

Ehrenfest-Afanassjewa, T. (1924b). Moet het meetkundeonderwijs gewijzigd worden? *Bijvoegsel van het Nieuw Tijdschrift voor Wiskunde*, *1*(2), 47–59. https://archief.vakbladeuclides.nl/bestanden/001_1924-25_02.pdf.

Ehrenfest-Afanassjewa, T. (1931). *Uebungensammlung zu einer geometrischen Propädeuse*. The Hague: Martinus Nijhoff, English translation by Hoechsmann (2011).

Ehrenfest-Afanassjewa, T., & Freudenthal, H. (1951). *Kan het wiskundeonderwijs tot de opvoeding van het denkvermogen bijdragen?* Purmerend.

José Ferreirós, J. (2007). *Labyrinth of thought: A history of set theory and its role in modern mathematics* (2nd ed.). Birkhäuser.

Furinghetti, F., Matos, J. M., & Menghini, M. (2013). From mathematics and education, to mathematics education. In M. A. (Ken) Clements, A. J. Bishop, C. Keitel, J. Kilpatrick, & F. K. S. Leung, (Eds.), *Third international handbook of mathematics education* (Vol. 27, pp. 273–302). Springer international handbooks of education. New York: Springer. https://doi.org/10.1007/978-1-4614-4684-2.

Giaquinto, M. (2002). *The search for certainty: A philosophical account of the foundations of mathematics*. Oxford: Oxford University Press.

Glas, E. (2002). Klein's model of mathematical creativity. *Science & Education*, *11*(1), 95–104. https://doi.org/10.1023/A:1013075819948.

Grattan-Guinness, I. (2000). *The search for mathematical roots, 1870–1940: Logics, set theories and the foundations of mathematics from cantor through Russell to Gödel*. Princeton and Oxford: Princeton University Press.

Gray, J. (2008). *Plato's Ghost: The modernist transformation of mathematics*. Princeton and Oxford: Princeton University Press.

Gray, J. (2010). *Worlds out of nothing: A course in the history of geometry in the 19th century*. Springer undergraduate mathematics series. Berlin: Springer. https://doi.org/10.1007/978-0-85729-060-1.

Gray, J. (2019). Epistemology of geometry. In E. N. Zalta (Ed.), *The Stanford encyclopedia of philosophy*. Metaphysics Research Lab, Stanford University, fall 2019 edition.

Gray, J. J. (1992). Poincaré and Klein—groups and geometries. In Boi et al. (Eds.) (1992, pp. 35–44). https://doi.org/10.1007/3-540-55408-4_51.

Hallett, M., & Majer, U. (Eds.). (2004). *David Hilbert's lectures on the foundations of geometry 1891–1902*. Berlin, Heidelberg: Springer.

Heinzmann, G. (2013). *L'Intuition Épistémique*. Vrin.

Heinzmann, G., & Stump, D. (2017). Henri Poincaré. In E. N. Zalta (Ed.), *The Stanford encyclopedia of philosophy*. Metaphysics Research Lab, Stanford University, winter 2017 edition.

Hilbert, D. (1899). *Grundlagen der Geometrie*. In B. G. Teubner, Leipzig (Eds.), *The foundations of geometry* (2004, pp. 436–525). Reprinted in Hallett and Majer (2004, pp. 436–525). Translated into English as *The Foundations of Geometry*, Open Court, Chicago, 1902.

Hoechsmann, K. (2011). Revisiting Tatjana Ehrenfest-Afanassjewa's (1931) "Uebungensammlung zu einer geometrischen Propädeuse": A translation and interpretation. *The Montana Mathematics Enthusiast*, *8*(1), 113–146. https://scholarworks.umt.edu/tme/vol8/iss1/6.

Klein, F. (1872). *"Vergleichende Betrachtungen ueber neuere geometrische Forschungen", Programm zu Eintritt in die philosophische Fakultät und den Senat der K. Friedrich-Alexanders-Universität zu Erlangen*. Erlangen: Deichert.

Klein, F. (1893). On the mathematical character of space-intuition and the relation of pure mathematics to the applied sciences. In R. Fricke, & H. Vermeil (Eds.), *Gesammelte mathematische Abhandlungen, Anschauliche Geometrie Substitutionsgruppen und Gleichungstheorie zur Mathematischen Physik* (Vol. II, pp. 225–231). Berlin, Heidelberg: Springer.

Klein, F. (2016). *Elementary mathematics from a higher standpoint* (Vol. II). Geometry. Berlin, Heidelberg: Springer. https://doi.org/10.1007/978-3-662-49445-5. English translation by Gert Schubring.

La Bastide-van Gemert, S. (2015). *All positive action starts with criticism: Hans Freudenthal and the didactics of mathematics*. Springer. Translation from the Dutch language edition "Elke positieve actie begint met critiek": Hans Freudenthal en de didactiek van de wiskunde, by Sacha La Bastide-van Gemert (2006). Hilversum: Verloren.

Mattheis, M. (2019). Aspects of "Anschauung" in the work of Felix Klein. In Weigand et al. (Eds.), (2019, pp. 93–106). https://doi.org/10.1007/978-3-319-99386-7_7.

Meikle, L. I., & Fleuriot, J. D. (2003). Formalizing Hilbert's Grundlagen in Isabelle/Isar. In D. Basin, & B. Wolff (Eds.), *Theorem proving in higher order logics* (pp. 319–334). Berlin, Heidelberg: Springer. https://doi.org/10.1007/10930755_21.

Nabonnand, P. (2007). Les réformes de l'enseignement des mathématiques au début du XXe siécle. Une dynamique à l'échelle international. In H. Gispert, N. Hulin, & C. Robic (Eds.), *Sciences et enseignement. L'exemple de la grande réforme des programmes du lycée au début du XXe siècle* (pp. 293–314). Vuibert. https://hal.archives-ouvertes.fr/hal-01083143.

Pasch, M. (1882). *Vorlesungen über neuere geometrie*. Leipzig: B.G. Teubner.

Poincaré, H. (1902). *La science et L'Hypothèse*. Flammarion, English translation in English translation in Poincaré (1913).

Poincaré, H. (1905). *Le Valeur de la science*. Flammarion, English translation in Poincaré (1913).

Poincaré, H. (1908). *Science et methode*. Flammarion, English translation in Poincaré (1913).

Poincaré, H. (1913). *The foundations of science: Science and hypothesis, the value of science, science and method* (Vol. 1), Science and education. The Science Press, English translation by G. B. Halstead of Poincaré (1902), (1905), (1908).

Poincaré, H. (1958). *The value of science*. Dover, Republication of the English translation by G. B. Halstead of Poincaré (1905)

Rowe, D. E. (1983). A forgotten chapter in the history of Felix Klein's Erlanger Programm. *Historia Mathematica, 10*(4), 448–454. https://doi.org/10.1016/0315-0860(83)90006-X.

Rowe, D. E. (1985). Felix Klein's "Erlanger Antrittsrede": A transcription with English translation and commentary. *Historia Mathematica, 12*(2), 123–141. https://doi.org/10.1016/0315-0860(85)90003-5.

Rowe, D. E. (1989). Klein, Hilbert, and the Göttingen mathematical tradition. *Osiris, 5*, 186–213. https://doi.org/10.2307/301797.

Rowe, D. E. (1992). Klein, Lie, and the "Erlanger Programm". In Boi et al. (Eds.) (1992, pp. 45–54). https://doi.org/10.1007/3-540-55408-4_52.

Rowe, D. E. (1994). The philosophical views of Klein and Hilbert. In Chikara, S., Mitsuo, S., & Dauben, J. W. (Eds.), *The intersection of history and mathematics* (Vol. 15, pp. 187–202), Science Networks · Historical studies. Birkhäuser, Basel. https://doi.org/10.1007/978-3-0348-7521-9_13.

Rowe, D. E. (2013). Mathematical models as artefacts for research: Felix Klein and the case of Kummer surfaces. *Mathematische Semesterberichte, 60*(1), 1–24. https://doi.org/10.1007/s00591-013-0119-8.

Schlimm, D. (2010). Pasch's philosophy of mathematics. *The Review of Symbolic Logic, 3*(1), 93–118. https://doi.org/10.1017/S1755020309990311.

Schlimm, D. (2013). The correspondence between Moritz Pasch and Felix Klein. *Historia Mathematica, 40*(2), 183–202. https://doi.org/10.1016/j.hm.2013.02.001.

Sinaçeur, H. (1993). Du formalisme à la constructivité: Le finitisme. *Revue Internationale de Philosophie, 47 186*(4), 51–283.

Smid, H. J. (2009). Foreign influences on Dutch mathematics teaching. In K. Bjarnadóttir, F. Furinghetti, & G. Schubring (Eds.), *"Dig where you stand". Proceedings of the conference on on-going research in the history of mathematics education* (pp. 209–222), Reykjavik: University of Iceland, School of Education.

Smid, H. J. (2012). The first international reform movement and its failure in the Netherlands. In K. Bjarnadóttir, F. Furinghetti, J. Manuel Matos, & G. Schubring (Eds.), *"Dig where you stand" 2. Proceedings of the second "International conference on the history of mathematics education", October 2–5, 2011, New University of Lisbon, Portugal* (pp. 463–476). UIED, Unidade de Investigação, Educação e Desenvolvimento, Lisbon.

Smid, H. J. (2016). Formative years: Hans Freudenthal in prewar Amsterdam. In *History and pedagogy of mathematics*. France: Montpellier. https://hal.archives-ouvertes.fr/hal-01349232.

Tobies, R. (2000). In spite of male culture: Women in mathematics. In R. Camina, & L. Fajstrup (Eds.), *European women in mathematics. Proceedings of the ninth general meeting, Loccum, Germany, 30 August–4 September 1999* (pp. 25–35). Hindawi. http://downloads.hindawi.com/books/9789775945020.pdf.

Tobies, R. (2012). The developent of Göttingen into the Prussian centre of mathematics and the exact sciences. In N. A. Rupke (Ed.), *Göttingen and the development of the natural sciences*. Wallstein.

Tobies, R. (2019). Felix Klein–Mathematician, academic organizer, educational reformer. In Weigand et al. (Eds.) (2019, pp. 5–21). https://doi.org/10.1007/978-3-319-99386-7.

Tobies, R. (2020). Internationality: Women in Felix Kleins Courses at the University of Gottingen (1893–1920). In E. Kaufholz-Soldat & N.M.R. Oswald (Eds.), Against All Odds: Womens Ways to Mathematical Research Since 1800 (2020, 9–38). Cham: Springer.

Torretti, R. (1978). *Philosophy of geometry from Riemann to Poincaré*. Dordrecht: D. Reidel.

van Hiele, P. .M. (1975). Freudenthal en de didaktiek der wiskunde. *Euclides, 51*(1): 8–10. https://archief.vakbladeuclides.nl/bestanden/051_1975-76_01.pdf.

Weigand, H. G., McCallum, W., Menghini, M., Neubrand, M., & Schubring, G. (Eds.) (2019). *The legacy of Felix Klein*. ICME-13 Monographs. Springer, Cham. https://doi.org/10.1007/978-3-319-99386-7.

Chapter 3
Afanassjewa and the Foundations of Thermodynamics

Jos Uffink and Giovanni Valente

Abstract We review aspects of Afanassjewa's work on the foundations of thermodynamics from her 1925 paper on the Second Law and her 1956 book *Grundlagen der Thermodynamik.* We argue that her work contained several valuable original insights in these foundations, often much ahead of her times. In particular, we discuss how her 1956 book anticipated and showed the way to solve an alleged paradox about reversible processes raised by Norton (2014, 2016) and discuss the remarkable comments in her 1925 paper on the asymmetry between work and heat exchange — which still awaits more common recognition—, and on the conceptual possibility of negative absolute temperatures, long before Ramsey (1956) made this an accepted physical possibility.

3.1 Introduction

Thermodynamics was borne out of the investigations of Sadi Carnot into the efficiency of heat engines in 1824. It developed in the second half of the nineteenth century as a mature and versatile physical theory through the work of Clausius, Kelvin, Planck, Gibbs and many others. The approach taken by most authors in this period was to develop this theory as a description of thermal properties of macroscopic physical systems, while staying aloof from any speculation about their microphysical constitution. In part, this methodology was favoured by the sceptical philosophical attitude against the atomic hypothesis, championed by authors like Mach; for another and probably more important part, it was favoured because proposals about the precise nature of the atomic constitution of matter in that period were mostly speculative and unsuccessful in explaining or predicting more than a handful of physical phenomena.

J. Uffink
Philosophy Department, University of Minnesota, Minneapolis, US
e-mail: uffink@umn.edu

G. Valente (✉)
Department of Mathematics, Politecnico di Milano, Milan, Italy
e-mail: giovanni.valente@polimi.it

J. Uffink et al. (eds.), *The Legacy of Tatjana Afanassjewa*, Women in the History of Philosophy and Sciences 7, https://doi.org/10.1007/978-3-030-47971-8_3

Today, one often regards this approach of the founding fathers of thermodynamics to avoid any assumption about the microphysical constitution of thermodynamic systems as a typical case of 'cold feet', especially after Einstein's work on Brownian motion and Perrin's (1913) *Les Atomes*. However, it did pay off handsomely by yielding a theory that, while our conceptions of microphysics have gone through revolutionary changes in the twentieth century with the advent of relativity and quantum mechanics, remains applicable even in areas widely beyond the typical cases studied in the nineteenth century (e.g. gases and liquids), like photon gases, magnets and spin glasses, and even black holes.

Thus, Einstein wrote in his autobiographical notes:

> A theory is the more impressive the greater the simplicity of its premises, the more different kinds of things it relates, and the more extended its area of applicability. Therefore the deep impression that classical thermodynamics made upon me. It is the only physical theory of universal content which I am convinced will never be overthrown, within the framework of applicability of its basic concepts. Einstein (1949)

Nevertheless, the formulation of thermodynamics that emerged in the Clausius-Kelvin-Planck tradition remained close to working intuitions of engineers and experimental physicists and does not meet the standards of rigour that would today be expected from a physical theory that aspires to such universal scope. The first mathematician who endeavoured to provide a more rigourous basis for thermodynamics was Carathéodory (1909). His work was mostly ignored by the contemporary physics community until Born (1921) published a sympathetic review (and much simplified version) of Carathéodory's paper Born (1921). Born's paper did attract a reaction from Planck and other physicists like Landé, and Ruark. Planck's assessment of Carathéodory's approach was, however, harsh: he called it an 'artificial and unnecessary complication' (Planck 1926).

Tatiana Afanassjewa (1925) also jumped into this discussion. Her response to Carathéodory was considerably more positive, while pointing out several lacunae in his treatment. What is more, she called for an even much more radical reformulation of thermodynamics, conceptually separating the treatment of thermodynamic equilibrium from that of irreversible processes. Indeed, she made clear in this paper that her own goal differed from Carathéodory in that she aimed to pursue a logical separation between those aspects of thermodynamics that deal purely with equilibrium states or the structure of the equilibrium state space of thermodynamical systems, and those aspects that deal with processes that such systems undergo in the course of time. Afanassjewa published several other articles on foundations of thermodynamics in Russian in 1928 and 1930, and two papers in Dutch in *Wis-en Natuurkundig Tijdschrift* (1936a, 1936b) (with a response by Verschaffelt (2019)).[1] She wrote a

[1] She also submitted a manuscript in French to *Actualités Scientifiques et Industrielles* in 1941. Unfortunately, it seems this latter manuscript was not received, presumably because of the war; when she resent her manuscript to the same journal in 1946, it seems to have not received a timely procedure, judging from what she wrote in her Ehrenfest-Afanassjewa (1948):

substantial manuscript in the 1940s[2] finally published as *Grundlagen der Thermodynamik* in 1956, when she was 80 years of age. But clearly much of the contents of this work was written much earlier.

Even though Afanassjewa's manuscripts are historically to be placed in relation to the papers by Carathéodory and Born, many of the pertinent points she made along the way remain truly worthwhile today, even when one disregards the historical context of Carathéodory's axiomatic approach (which has in fact been superseded by more recent rigourous approaches to thermodynamics, in particular by Lieb and Yngvason (1999)). The purpose of this paper is to point out and review these aspects of her work and show their relevance to a very recent debate in the foundations of thermodynamics on an alleged 'paradox of reversible processes'.

Before dwelling upon these issues, we like to add a few remarks to emphasize the extent of Afanassjewa's influence. As reported by Van der Heijden elsewhere in this volume, Paul Ehrenfest and Tatiana Afanassjewa influenced many young students in Leiden that regularly met for informal discussions in their home, which she had designed particularly for this purpose. The foremost influences of Afanassjewa's views on thermodynamics on this cohort of young academics are on Philipp Kohnstamm, and Berta Lorentz, (H.A. Lorentz's oldest daughter). Kohnstamm had been a student of J.D. Van der Waals, and co-authored their textbook *Lehrbuch der Thermodynamik* in 1908. A third edition of this book appeared in 1927. By this time, Van der Waals had died and Kohnstamm took the revision upon himself. He explains in the foreword to this edition that he changed the presentation considerably, and even changed the title of the book, which now appeared as *Lehrbuch der Thermostatik* (Van der Waals and Kohnstamm 1927), to underline the conceptual separation of considerations of equilibrium states (statics) from those of processes. He notes in this foreword 'I am much endebted to T. Ehrenfest-Afanassjewa in reaching full clarity on this point' and frequently refers to oral or written discussions with her in many sections of the book. Berta Lorentz (or De Haas-Lorentz, after her marriage to the physicist Wander de Haas) wrote a Ph.D. thesis on Brownian motion, and co-authored with Afanassjewa a paper on the Le Chatelier-Braun principle in thermodynamics. Of particular relevance for our purpose is the textbook on thermodynamics she wrote *Over de beide hoofdwetten der thermodynamica en toepassingen* in 1939, that, like Kohnstamm's, was written from Afanassjewa's point of view.

These examples attest to Tatiana Afanassjewa's influence on a generation of physicists that contributed to the development of thermodynamics in the first part of the

> In the year 1941, I sent my paper "*Le second Principe de la Thermodynamique et l'Irreversibilité*" to the editors of *Actualités Scientifiques et Industrielles* .... At that time, my mailing did not reach the editors. Now, the paper finally lies with the editors –for more than a year.

Yet, this paper was never published.

[2]A version of this manuscript (1948), as lecture notes for a course in the year 1947/48, is preserved in Leiden. A letter from her daughter Tanya from January 1941, with detailed comments on the manuscript, (which one of us (J.U) was allowed to read by Afanassjewa's granddaughter Tamara van Bommel) show that an even earlier version of this manuscript must have been written in 1940.

20th century, who had the privilege of directly interacting or just corresponding with her. Yet, as we explain in this paper, her legacy extends much beyond her contemporaries.

The organization of this paper is as follows. We begin in the next section by placing her work within the mathematical tradition to axiomatize thermodynamics starting with Carathéodory. In Sect. 3.3 we discuss her (1956) distinction between quasi-static processes and quasi-processes and explain how that helps one elude an ostensive paradox that has engendered a recent debate in the philosophical literature. In Sects. 3.4 and 3.5 we focus on two remarkable topics of her (1925) paper. In Sect. 3.4 we discuss her analysis of the question whether the differential equation for heat and work should be treated on a par, or whether there is a crucial asymmetry between them not captured in other approaches to thermodynamics. Finally, in Sect. 3.5 we focus on Afanassjewa's analysis of the Second Law of thermodynamics: remarkably, she was the first author to point out that the alleged equivalence between Clausius' statement and Kelvin's statement really rests on the assumption that absolute temperature always be positive, rather than negative. We argue that, by contemplating the conceptual possibility of negative absolute temperatures, she was 30 years ahead of her time.

3.2 The Axiomatic Approach to Thermodynamics: From Carathéodory to Afanassjewa

Carathéodory (1909) made the first attempt to axiomatize thermodynamics as a mathematically rigourous theory. Carathéodory had before been working on areas which we would now call topology, which was still at a fledgling stage at this time. His main achievement, looking back with hindsight, is to formulate thermodynamics as a theory about the state space Γ_{eq} of all equilibrium states of a thermodynamical system, and to assume this space forms a differential manifold. That is to say, the space Γ_{eq} is a mathematical construct which is itself 'coordinate-free'; coordinates are only introduced to chart the manifold, and the main condition to guide our choice of coordinates is that they disambiguate the states (and some choices of coordinates may fail this condition, even for a fluid as familiar as water, cf. Thomsen and Hartka (1962)).

So, suppose we have a thermodynamics system in an equilibrium state, represented as a point $s \in \Gamma_{eq}$, and that there is suitable choice of coordinates for this space, in which we can characterize this point by its coordinates as $s = (x_1, \ldots x_n)$. There are supposed to be (at least) two independent ways of interacting with a thermodynamic system: by exchanging heat Q or by doing work W. Let us suppose (for the purpose of this section, somewhat uncritically) that processes, in which the system exchanges heat with or does work on its environment (or both), can be made to proceed so slowly compared to relevant equilibration processes that during the interaction one can still at all times (or up to a negligible error) characterize its state by a point in Γ_{eq}. The change of state during such a process will then be represented by a smooth curve in Γ_{eq}. The nomenclature of such processes varies among authors: Planck called them 'infinitely slow' or 'reversible' processes, Carathéodory called them 'quasi-static changes of state', Born called them 'reversible processes', and Afanassjewa (1925) 'quasi-static processes'. We will employ this last name here.

If we focus on an infinitesimal section of such a curve, the effect of the interaction on the state of the system may be given by some differential equation, like

$$đQ = \sum_{i=1}^{n} X_i \mathrm{d}x_i \tag{3.1}$$

for heat exchange; and for an exchange of work by a similar expression:

$$đW = \sum_{i=1}^{n} Y_i \mathrm{d}x_i \tag{3.2}$$

where the X_i and Y_i are assumed to be smooth functions on Γ_{eq}, which will depend on the kind of system. For example, if the system is one mole of an ideal gas, whose state space is two dimensional, and if we choose the coordinates as (T, V) (its absolute temperature and volume), the expression (3.1) becomes

$$đQ = sc_V \mathrm{d}T + p\mathrm{d}V. \tag{3.3}$$

where c_V denotes its specific heat at constant volume, an $p = RT/V$ its pressure. Similarly, the work done by the system in this case will be expressed as

$$đW = p\mathrm{d}V. \tag{3.4}$$

But for other systems or indeed for different choices of coordinates the functions X_i in (3.1) or Y_i in (3.2) will need to be adapted accordingly.

Of special interest are the so-called *adiabatic* quasi-static processes, in which there is no heat exchange with the system, and therefore, Eq. (3.1) becomes

$$đQ = \sum_{i=1}^{n} X_i \mathrm{d}x_i = 0. \tag{3.5}$$

A central question is then whether or not there exists a quantity (which will eventually get the name of 'entropy' S) such that adiabatic quasi-static processes have the property that S remains constant in any such process. In other words, the question is whether there exists a function S on Γ_{eq} such that the curves that represent adiabatic quasi-static processes will always lie in a hypersurface characterized by the equation

$$S(x_1, \ldots, x_n) = \text{const.} \tag{3.6}$$

This is not a trivial question, especially not when the dimension of Γ_{eq} is higher than two.

Another notable achievement of Carathéodory is that he drew a connection between this question and the work by Pfaff from the 19th century, that showed that a partial differential equation like (3.5) could belong to exactly three different classes:

(i) It might be that there exists some function Q on Γ_{eq}, such that

$$\frac{\partial Q}{\partial x_i} = X_i \text{ for } i = 1, \ldots, \text{n} . \tag{3.7}$$

In that case, Eq. (3.5) is simple: it boils down to $dQ = 0$, and curves that obey it will indeed belong to the hypersurface $Q =$ const. In this case, Eq. (3.5) is called integrable, and dQ is said to be an exact differential.

(ii) It might also be that there does not exist such a function Q as imagined in case (i). In this case, $\text{đ}Q$ is not integrable, and called an *inexact* differential (which motivates the notational distinction $\text{đ}Q$ instead of dQ). However, a weaker condition will nevertheless yield similar implications as in case (i). This condition is the assumption that there are two functions on Γ_{eq}, say T and S, where $T \neq 0$ everywhere, such that

$$X_i = T\frac{\partial S}{\partial x_i} \text{ for } i = 1, \ldots, \text{n} \tag{3.8}$$

Under this condition Eq. (3.5) is equivalent to

$$TdS = 0, \tag{3.9}$$

which in view of the assumption that T is non-vanishing, is equivalent to $dS = 0$, by which we recover the same implications as in case (i), this time for the hypersurface $S =$ const. If this condition holds, it is said that there exists an 'integrating divisor' for Eq. (3.5), since a division of the equation on both sides by T makes it integrable.

(iii) It might also be that both cases just considered fail to hold. Indeed, it could be that a smooth curve, starting out at any given initial point $s_{in} \in \Gamma_{eq}$, always obeying Eq. (3.5), we can reach any other desired point in Γ_{eq}. In such a case, Eq. (3.5) simply doesn't constrain a curve obeying this equation to any hypersurface at all.

Now, it is well-known that the adiabatic quasi-static processes in thermodynamics do not belong to class (i), even for a case as simple as the ideal gas. The physical interpretation of this fact is that heat Q is not a quantity, i.e. one cannot characterize systems by the 'amount of heat' they contain, as was supposed by many 18th and early 19th century authors. On the other hand, if the adiabatic quasi-static processes in thermodynamics would belong to class (iii), it would make most of that theory utterly inapplicable. It is therefore important that all such processes belong to class (ii), i.e. that one can guarantee the mathematical existence of an integrating divisor, namely, the temperature T, such that division of the inexact differential $\text{đ}Q$ by T turns it into the exact differential $dS = \text{đ}Q/T$.

Carathéodory also proposed a particular proposition (Carathéodory's Principle) that he claimed to imply that adiabatic quasi-static processes belong to class (ii), and would therefore guarantee the existence of entropy for a thermodynamical system, and in his view express the Second Law of thermodynamics.[3]

Afanassjewa, being trained as a mathematician herself, applauded Carathéodory's work in her (1925, p. 933) as 'a particularly valuable attempt to axiomatize the Second Law of thermodynamics'. However, she set herself the task of elaborating a more complete and conceptually quite distinct axiomatic formulation of the theory, while building upon Carathéodory's results. As she put it in the opening lines of her 1925 paper:

> [T]he approach here is different from Carathéodory's, most of all concerning the role of irreversible processes. The aim here is to obtain a deeper understanding of the Second Law by concentrating exclusively on reversible processes. (p. 933.)

Indeed, her paper presents two sets of logically independent axioms, the first set aiming to describe the properties of thermal systems in equilibrium and the second set aiming to characterize the notion of irreversibility in contrast with the properties of reversible or quasi-static processes, conceived of as curves in equilibrium space. This approach culminated in her 1956 book, where she put forwards a more careful characterization of the concept of thermodynamical processes, which eliminates the use of the misleading adjective 'reversible'. In the next section, we discuss this issue in greater detail.

3.3 Reversible Processes Versus Quasi-Processes

As we have seen, a discussion of 'reversible processes' (or 'quasi-static processes 'infinitely slow processes', etc.) is central to thermodynamics. In the traditional understanding, such processes are represented by curves in the equilibrium state space of a given thermodynamic system. In Afanassjewa's 1925 paper, she endorsed this view, following Carathéodory. However, in her 1956 book, she developed a more subtle viewpoint, in which such curves are not taken to be representative of processes at all but seen merely as 'quasi-processes'. In this section, we aim to explain her reasons for introducing this new terminology and show why it matters.

Representing reversible processes as curves in equilibrium space has the immediate advantage that it allows application of the tools of calculus to characterize such curves locally by means of partial differential equations. In particular, one can write

[3] More details on the formulation of Caratheodory's Principle, and on the question whether his claim holds, see (Uffink 2001).

the differential form of the First Law as $đQ = dU + đW$, where $đQ$ represents an infinitesimal quantity of heat added to the system in a reversible process, dU the change in internal energy and $đW$ the work done by the system against its environment. Afanassjewa's *Axiom A* then asserts that, if the integral $\int_1^2 đQ$ is non-zero along some curve connecting two equilibrium states 1 and 2, there is no adiabatic curve between these two states. This entails Carathéodory's principle, from which one can in turn prove the existence of an entropy function S such that $đQ = TdS$, by means of the Pfaff theorem described above. Yet, Afanassjewa went on to observe that, contrary to what Carathédory claimed, this fact is not sufficient to establish the Second Law, for which she introduced further axioms that we will address in later sections below.

What we want to address first is that the conception of reversible processes as curves in equilibrium space raises the threat of a paradox that has been explicitly formulated recently by Norton (2014, 2016). Roughly speaking, the paradox arises because a system in a thermodynamic equilibrium state does not change its state in the course of time as long as it remains isolated. To make a system undergo a *process*, i.e. to change its state, we need to disturb it from outside, either by, e.g. removing a partition, by exchanging heat or work with it, etc. But when the system is disturbed, it will no longer be in equilibrium, and its state cannot be described in the equilibrium state space. This raises the question of whether curves in the equilibrium state space of a system can be understood as processes at all. Norton formulates this paradox as a contradiction between two claims about reversible processes (using a view in which this disturbance is characterized as 'an imbalance of driving forces'):

1. They are processes with a non-equilibrium imbalance of driving forces, such as non-zero temperature differences or unbalanced mechanical forces; for this imbalance is needed to move the system from one state to another.
2. At the same time they are sets of equilibrium states in which, by definition, there is no imbalance of forces; for then the forwards and the reverse processes pass through the same set of equilibrium states and both can be represented by the same curve in equilibrium state space. [Norton (2016), p. 43]

Norton submitted that this paradox is at the bottom of a mistake that he claimed to be common to virtually all authors in thermodynamics, including Afanassjewa herself. In his view, the alleged mistake rests on the interpretation of reversible processes as proper idealizations, namely, as consistently defined infinitely long processes, and it can be avoided if one uses curves of equilibrium states simply as yielding convenient approximations of the properties of extremely slow real thermodynamical processes. Nevertheless, Valente (**?**) objected that such a characterization was already implicit in Afanassjewa's work and showed that one can disarm Norton's paradox of reversible processes by resorting to her notion of quasi-processes. Here we elaborate on her proposal.

Indeed, the most natural step to prevent Norton's paradox from arising is to recognize that curves in equilibrium state space are not to be understood literally as descriptions of *processes* (i.e. as changes of state of a system occurring in the course of time) at all. It is this recognition that led Afanassjewa, in her 1956 book, to take the crucial step of denoting such curves as '*quasi-processes*', rejecting all earlier terminology, whether it was the traditional term 'reversible process', Carathéodory's 'quasi-static change of state" or Planck's 'infinitely slow process'.

In the opening chapter of her 1956 book Afanassjewa noted that if a system is initially in equilibrium, any process requires an external disturbance, by which the system attains a non-equilibrium state (Sect. 3.5). In that section, she discussed the simplest example of such a process in which the disturbance is just a one-time external intervention: the removal of a partition or loosening of an internal piston; or the establishment of a new coupling between previously separated systems (say: a thermal coupling between bodies at different temperatures, or a mechanical coupling between bodies at different pressure, etc.). She called those processes, where a single external intervention introduces a non-equilibrium condition, and the system is afterwards left to evolve in isolation *spontaneous processes*. In the following section (Sect. 3.6) she discussed the state-space representation of such processes, by means of an extremely high-dimensional manifold (which we call Γ_{neq}) containing the non-equilibrium states of the system, in which the space of equilibrium states Γ_{eq} is imbedded as a very low-dimensional subspace (*Diagonalraum*).

A process in a thermodynamical system, even if it starts in an equilibrium state in Γ_{eq}, will thus generally be represented by curve in Γ_{neq} that immediately takes it out of this equilibrium subspace. Under certain conditions, one may assume that the system will eventually evolve towards a new equilibrium state.[4] In Sect. 8, spontaneous processes are contrasted with what Afanassjewa called 'forced changes of states', wherein disturbances from equilibrium are not caused by the removal or introduction of couplings inside a system, which modify its internal structure, but rather by a coupling of intensive parameters of the system with a changing environment. In particular, she focussed on a special class of those forced changes of states during which (i) one or more of the intensive equilibrium parameters (i.e. temperature or pressure, etc) of the system and of its environment differ only slightly so that (ii) the system and its environment remain close to obeying the conditions of equilibrium; (iii) the difference between the final and initial values of the parameter for the system is also very small. As a result, the average velocity of equilibration of the parameters will also be very small. She calls such a change of state an *elementary quasi-static process*. Its state-space representation is by a curve in Γ_{neq} whose initial and final states are in the subspace Γ_{eq}, while all its other points are outside but near to this subspace.

She then defines a complete quasi-static process as a succession of a large number of such elementary quasi-static processes.[5] Implicit in this conception is that we also apply a limiting procedure, in which the parameter difference between system and environment tends to zero in each elementary quasi-static process, while the number

[4]She discusses a counterexample for these 'certain conditions' by means of a gas expanding into an infinite volume. In (Brown and Uffink 2001), this claim that an isolated thermodynamic system eventually reaches equilibrium is codified as the 'Minus First Law', under the condition that the system is contained in a fixed finite volume.

[5]Note that for Afanassjewa, quasi-static processes are always forced processes; a sequence of spontaneous processes, even if they are made to proceed very slowly, is not.

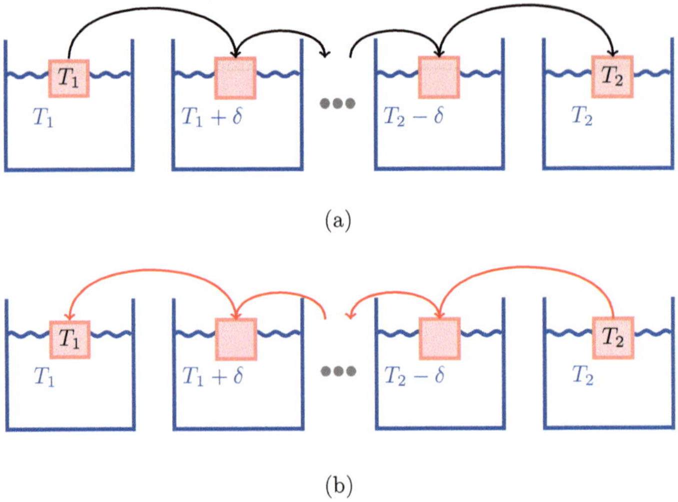

Fig. 3.1 **a** An illustration of how Afanassjewa conceived of a quasi-static process for raising the temperature of a system from T_1 to T_2 (where $T_2 = T_1 + N\delta$). As a starting point, the system is in thermal equilibrium with a heat reservoir at an in initial temperature T_1. Then, we move the system to another heat reservoir at a slightly higher temperature $T_1 + \delta$, and let it equilibrate with this reservoir. During this process of equilibration, the system will *not* by in equilibrium at all, but after some period, it will reach a new equilibrium at the temperature $T_1 + \delta$. (This completes an 'elementary quasi-static process'.) Repeat this step by now placing the system in a heat reservoir at temperature $T_1 + 2\delta$, etc., until after N steps the system finally reaches equilibrium at T_2. Afanassjewa's conceived of a quasi-static process as a limiting procedure of such a sequence, in which the temperature steps δ tends to zero, while the number N of elementary steps to reach temperature T_2 grows to infinity. **b** A similar quasi-static process for lowering the temperature from T_2 to T_1. Starting from the right-hand side of the figure, where the system is in thermal equilibrium with the reservoir at temperature T_2, one constructs a sequence of elementary quasi-static processes by successively placing the system in heat baths that are each at a temperature δ *lower* than that of the preceding bath. The resulting quasi-static process is therefore *not* the exact reversal of the process in Fig. 3.1a

of such elementary processes goes to infinity. These ideas are illustrated in Fig. 3.1a. In a state-space representation, a quasi-static process thus corresponds to a discrete sequence of equilibrium points, joined by intermediate curves of non-equilibrium states (cfr. Fig. 3.2). Thus, Norton's paradox does not arise for such processes: in agreement with Statement 1 the system moves from state to state by non-equilibrium processes. However, since it is not represented by a continuous curve in equilibrium state space, Statement 2 is false.

For the purpose of characterizing equilibrium curves, Afanassjewa (1956) introduced the term *quasi-processes*, in order to distinguish it from the notion of quasi-static processes. The intended definition and the role of quasi-processes in thermodynamics are explained in the subsequent Sect. 9:

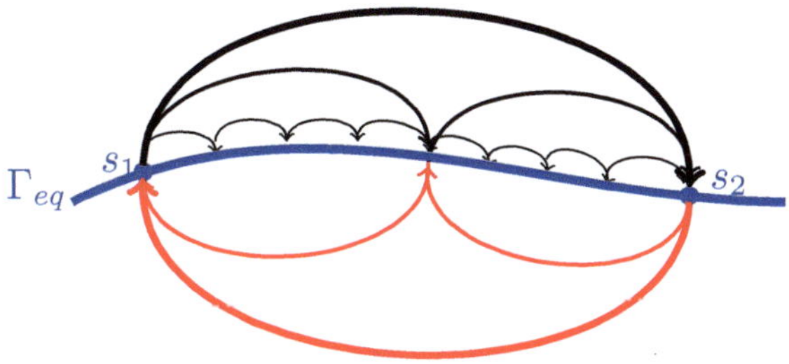

Fig. 3.2 A representation of the quasi-static processes of Fig. 3.1 in state space. Here, the blue curve represents the equilibrium subspace Γ_{eq} with two designated equilibrium states s_1 and s_2. The non-equilibrium processes in the upper half of the figure illustrate how one can change the state of the system from s_1 to s_2 by a sequence of processes in which the process is divided into smaller and smaller elementary quasi-static processes, a procedure by which they converge upon a continuous curve (quasi-process) within Γ_{eq}. The processes in the lower half illustrate the same for going from s_2 to s_1, showing that while none of the processes in the lower half are an exact reversal of processes in the upper half, they may converge onto the *same* quasi-process in Γ_{eq}. Indeed, while each of the approximating processes in either the upper half or lower half are generally irreversible, the limit to which they both converge is a quasi-process, for which the distinction 'reversible/irreversible' no longer makes sense

> [C]ontinuous sequences of equilibrium states that join two given states of equilibrium ...are represented graphically by curves in [Γ_{eq}]. They have traditionally been called "processes", and in particular "reversible" processes. We will rather call them "*quasi-processes*," as – clearly – they cannot be actualized by any real process; further, we explicitly want to leave out the epithet "reversible". (p. 13)
>
> Since quasi-processes are not processes, there is, therefore, nothing in them that could be reversed — except, perhaps, that one might say one could follow them mentally in both directions. But that is also not forbidden for any other series of non-equilibrium states a system goes through in a real process. If, on the other hand, one wants to consider the approximating quasi-static process, rather than the quasi-process itself, then those are irreversible in the same sense as any other real process. Accordingly, a quasi-static process by which one would want to approximate a quasi-process is not exactly the inverse of that by which one approximates the reverse quasi-process: if, for example, the direct quasi-process requires a continual flow of heat into the given system, then the systems in the environment which are used for this purpose must always have a higher temperature than the system during the direct course, while they should always be at a lower temperature during the reversed course." (pp. 15–16)

The last remarks in this quote are illustrated in Fig. 3.1b, which displays a reversal of the quasi-static process of Fig. 3.1a.

To appreciate the importance of Afanassjewa's proposal two remarks are in order. First of all, she clearly indicated that continuous curves in equilibrium space cannot themselves be interpreted as thermodynamical processes, and that is why she proposed that they should be better called quasi-processes. In fact, they are mere mathematical constructs that enable the application of partial differential equations to thermodynamics. As a consequence, quasi-processes naturally satisfy the property of being represented by curves in equilibrium space in Statement 2 of Norton's paradox, but they do not obey Statement 1, thereby again avoiding contradiction.

The connection between quasi-processes with real thermodynamical processes is instead enforced by the notion of a quasi-static limit. As we have seen, quasi-static processes are described as discrete sequences of equilibrium states, joined by non-equilibrium curves. In the quasi-static limit, where the number of elementary quasi-static processes grows, this discrete sequence becomes dense, and the intermediate non-equilibrium curves get shorter and closer to equilibrium space. One may argue that they will converge to a continuous curve in equilibrium state space. If so, and if a process that is sufficiently close to a quasi-process at each stage of its evolution, such a process may be regarded as practically indistinguishable from a quasi-process.[6]

The second remark concerns reversibility. As explicitly asserted in the above quote, Afanassjewa's nomenclature aims to eliminate the adjective 'reversible' from the characterization of equilibrium curves. In her view, a thermodynamical process is reversible just in case its inverse process, during which the system retraces all the steps in the reversed order, can also occur in nature. However, since quasi-processes are not actual thermodynamical processes, they cannot possibly be reversible. The fact that continuous curves in equilibrium state space may be equivalently parametrized in two directions (with either an increasing or decreasing parameter) does not have anything to do with the temporal direction in which thermodynamical phenomena occur. The underlying intuition was already present in her 1925 paper when she commented on the Second Law:

> Whether a particular state can be reached adiabatically quasi-statically from a given one only depends on the coefficients of the Pfaff equation, and nothing else. So we conclude, in the first place, that for the existence of entropy it is irrelevant whether the quasi-static processes themselves are reversible or not (p. 942).

In other words, the existence of an integrating divisor for heat, which ensures the existence of entropy, has nothing to do with the direction of time at all. In the 1956 book she further developed this view by emphasizing that, even if a quasi-static process approximates a continuous curve oriented in one direction, the quasi-process approximating the curve in the opposite orientation would not be the inverse process of the original one, as illustrated in Fig. 3.2. Later, in the 1959 Preface to the English translation of the famous encyclopedia article on the foundations of statistical mechanics she co-authored with Paul Ehrenfest, Afanassjewa reinforced this conviction by warning that the existence of an integrating divisor is completely independent of the direction in time in which to thermodynamical processes develop.

[6]For completeness, let us mention that Norton (2016) casted doubts upon Afanassjewa's notion of 'closeness to equilibrium'. (See Valente (?) for a reply.) Clearly, to make such a notion precise one needs to equip Γ_{neq} with an appropriately chosen topology (or, stronger, with a metric or distance) by which one can qualitatively or quantitatively express 'closeness to equilibrium', and to show that this choice is relevant to physical practice in the sense that if this 'distance from equilibrium' is small enough, the state is observationally indistinguishable from equilibrium. This is, indeed, a formidable technical task since Γ_{neq} is generally a space of extremely high dimension, and states can deviate from equilibrium in a myriads of different ways. Admittedly, Afanassjewa did not address this formidable task, except by qualitative and intuitive remarks. Neither, we may add, has Norton. Indeed, his notion of an 'imbalance in driving forces', which refers (at least partly) to the environment rather than the system itself, does not readily translate into a topology on the non-equilibrium state space of the system.

To sum up this section, we conclude that Afanassjewa in 1956 already successfully showed how to resolve a paradox that Norton raised about the traditional understanding of reversible processes in (2014, 2016). There seems to be a slight difference in emphasis in their approaches, though: while Norton is worried about the sense in which 'reversible processes' are processes; Afanassjewa was perhaps more worried about the sense in which they are 'reversible'.

3.4 The Distinction Between Heat and Work

As we saw in the previous section, Afanassjewa distinguished between spontaneous processes and forced changes of state, depending on whether a thermal system remains isolated after an external intervention, or whether it remains coupled with its environment, respectively. It is only in the latter case that one can define quasi-static processes. Any approach to thermodynamics, rich enough to express the Second Law, must also account for (at least) two ways in which the system can interact with the external environment: either by an exchange of work (positive or negative), or by exchanging heat. While Carathéodory assumed that there is a distinction between these two types of interaction, he did not comment on what this distinction was, presumably because it is hard to capture mathematically. The unfortunate result is that his formalism is actually symmetrical under the interchange of the meaning of 'heat' and 'work' (*mutatis mutandi*). However, in her 1925 paper, Afanassjewa identified an asymmetry to distinguish between these two types of interactions and included it in one of her axioms.

There is a *prima facie* analogy between the treatment of heat Q and work W in thermodynamics. In a quasi-static change of state[7] in which the system exchanges heat with its environment, the heat absorbed by the system is given by the equation:

$$đQ = TdS, \tag{3.10}$$

whereas the equation for the work done by the system, likewise in a quasi-static change of state, reads

$$đW = pdV \tag{3.11}$$

This pair of equations thus suggests a symmetry in the theory: if one were to replace 'heat absorbed' by 'work done', temperature T by pressure p, and entropy S by volume V, the two equations transform into each other. In both cases, the inexact differential of a given quantity that is not a function on Γ_{eq} (heat Q or work W) is related, by means of an integrating divisor (temperature T or pressure p) to the exact differential of some other quantity (entropy S, or volume V resp.) that *is* a function

[7]Here, we stick to the phrase 'quasi-static change of state' employed by Ehrenfest-Afanassjewa (1925), although, from the perspective of her Ehrenfest-Afanassjewa (1956), the term 'quasi-process' would be more appropriate.

of state. This establishes a formal symmetry between relations (3.10) and (3.11). With an eye towards the Second Law, one may note that just like there is no way to decrease the entropy of a system unless it exchanges heat to its environment, it is also true that one cannot decrease the volume of a body of gas without doing work on it.

So, thermodynamics appears symmetrical with respect to the differential Eqs. (3.10) and (3.11), both in terms of their formal structure and in terms of how they apply to the relevant processes of heat exchange and gas expansion. Nevertheless, the analogy breaks down as one takes the Second Law into consideration. For, the standard textbooks on thermodynamics teach us how, starting from Eq. (3.10), we can arrive at the conclusion that for any adiabatic process (i.e. any process in which there is no heat exchange), whether quasi-static or not, we obtain the result that entropy can never decrease. The Second Law is thus presented in the form:

> The entropy S of a system is non-decreasing in any adiabatic process.

There is also a presumption that entropy is singled out as *the* measure of irreversibility. By contrast, standard textbooks on thermodynamics never present an analogue of the Second Law for the performance of work, which would be a statement of the following form:

> The volume V of a system is non-decreasing in any process in which no work is exchanged with the system.

The usual textbooks on thermodynamics typically discuss only two ways in which the volume of a gas can change: either by moving a piston against an external pressure, which always involves a non-zero exchange of work, $đW$; or by removing a partition and allowing the gas to expand in a vacuum to attain a larger volume, and equilibrate, this time without work, $W = 0$.[8] Yet, they contain no examples of processes in which the volume of a system could shrink without the performance of work. One might thus be led to expect that such processes are forbidden in thermodynamics.

To be sure, there actually are systems that can display states with negative pressure, such as elastic bands, certain fluids in metastable states, and more exotic examples in cosmological applications. Those systems will, indeed, spontaneously contract when their volume is allowed to change. But these are not amongst the staple example of elementary textbooks. So one might well ask what distinction between work and heat exchange implies an asymmetry between them.

Afanassjewa (1925) was, to our knowledge, the first author in the foundations of thermodynamics to point out that there actually is a conceptual distinction between processes involving heat exchange and the exchange of work, providing an explicit example for this purpose. Also, she captured the distinction between heat and work within her (1925) axiomatization of thermodynamics by means of an independent axiom, i.e. *Axiom B* or *Coupling axiom*. Let us discuss these two points in order.

In her 1925 paper Afanassjewa considered a system enclosed in a container with two movable pistons with the same area A (p. 936). The system is in mechanical

[8]Note, that according to Afanassjewa's account, the first process is an example of forced change of state, whereas the second is a spontaneous process, and therefore not quasi-static.

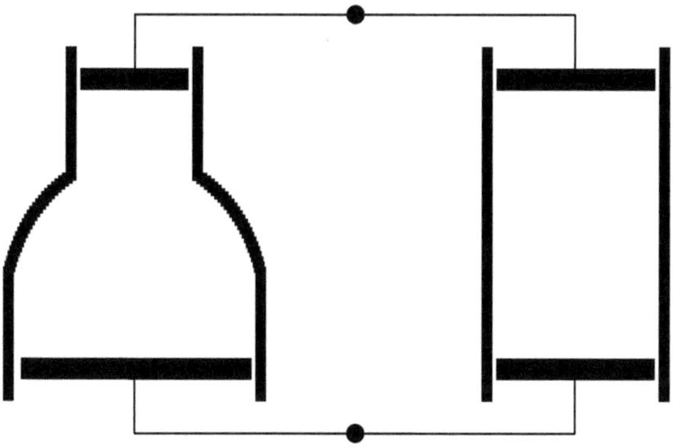

Fig. 3.3 Two vessels, each containing a body of gas are coupled mechanically, in this picture by means of equal-armed levers, initially assumed to be locked. By unlocking either the top or the lower lever, one can arrange that work will be done on system 1 by system 2, or alternatively, on system 2 by system 1, depending on their pressures and the areas of the enclosing pistons. This example shows that a thermodynamic system at low pressure can very well do work on another system with higher pressure (In contrast, a system with low temperature cannot transmit heat to system with high temperature

contact with a reservoir, considered as its environment, made of a gas enclosed in a container with two movable pistons of different areas A_1 and A_2, where the first is smaller than A and the second is larger (See Fig. 3.3). Suppose that initially all the pistons are locked in position (e.g. by pins through the walls), the pressure of the gas system is p and the pressure of the reservoir p_R is somewhere between $p\frac{A}{A_2}$ and $p\frac{A}{A_1}$. If we now unlock the upper coupling in Fig. 3.3, the system will do work on the reservoir, because the force that the system exerts on the lever is greater than the force by the reservoir on this lever. But if instead we release the lower lever, the reservoir will do work on the system. (In both cases, these will not be quasi-static processes, of course.) The point, here, is that the direction of work exchange is not determined by the values of the pressure p of the system and p_R of the reservoir. Since we assumed $p\frac{A}{A_2} < p_R < p\frac{A}{A_1}$ and $\frac{A}{A_2} < 1 < \frac{A}{A_1}$, p_R may be greater or smaller than p. Hence, this example shows that a system can very well perform work on another system with a higher pressure. The direction of the exchange of work depends on the particular mechanical contraption used to couple systems. (Of course, this should not come as a surprise, since many hydraulic devices exploit this feature.)

However, the point just made stands in marked contrast to the direction of heat exchange and its relation to the temperatures of the systems involved. Indeed, one of the classic formulations of the Second Law by Clausius (see Sect. 3.5 below) claims that the direction of heat exchange is *always* from a body at a higher temperature to a body a lower temperature. But if we replace 'heat' with 'work' and ' temperature' with 'pressure', the analogous statement is not true, as we have seen. According to

Afanassjewa, this marks a decisive distinction between the exchange of heat and work, or between thermal and mechanical coupling. Even though heat exchange can be effected by many different physical mechanisms (e.g. by radiation, conduction or convection), the question whether the system absorbs or emits heat from its environment depends only on their respective temperatures. She codified this point into the following independent axiom:

Axiom B. (*Coupling Axiom*) Only one kind of thermal coupling is possible.

De Haas-Lorentz (1938) phrased this axiom in the following words:

> There are no thermal levers.

Of course, both formulations of the coupling axiom may not have been the clearest use of words. We take it that these words are meant to express something like the following[9]:

> The direction of heat exchange in a thermal coupling between two bodies initially in equilibrium depends only on the temperatures of these bodies.

So, Afanassjewa 1925 recognized that there is an explicit asymmetry between heat exchange and work exchange. She convincingly argued, in the example she gave, that in contrast to the above, the direction in which work is exchanged by a mechanical coupling between two systems in equilibrium is not determined by their pressures but depends also on the kind of coupling: it is very well possible for a system with low pressure to do work on a system with a higher pressure.

For Afanassjewa, the Coupling axiom is motivated by the fact that its failure would lead to a violation the Second Law. Indeed, she presented it as a necessary condition for the derivation of the law in the sense that, if it did not hold, there would exist some closed curve in equilibrium space along which $\oint TdS \neq 0$ while extracting heat only from a single reservoir at constant temperature and convert it into work. In addition, she argued that, in combination with the previously discussed Axiom A and an additional Uniqueness Axiom C, which states that $\oint dS = 0$ along any closed path in Γ_{eq}.[10] Axiom B is sufficient in order to prove the important statement that more than one reservoir be needed if one wants to obtain work from heat during a thermodynamical cycle. Afanassjewa's sharp analysis on this point certainly helps to shed light onto the structure of the theory.

---

[9]We admit here that we are taking liberties with respect to Afanassjewa's writing. Her own paper expression the intention of the axiom as: 'What is meant, here, is that when two systems are coupled in such a way that they exchange heat, while maintaining equilibrium, this is only possible when they have the same temperature (rather than any other function of state)' (p. 935–936). This statement is indeed a consequence of our formulation, if we interpret the condition that the systems maintain equilibrium during the heat exchange the execution of a limit in which the heat exchange becomes quasi-static. The fact that she only mentioned this consequence for quasi-static processes of heat exchange might be due to her wish to strictly separate considerations of equilibrium from those of irreversibility.

[10]The need for Axiom C is somewhat unclear to the present authors, as it seems to us that $\oint dF = 0$ along a closed path for any continous function F on Γ_{eq}.

3.5 Afanassjewa on Negative Absolute Temperatures

There now remains one last step in Afanassjewa's axiomatic formulation of the Second Law of thermodynamics. Axioms A, B and C entail the existence of an entropy function S such that $đQ = TdS$. But one also needs to show that this function obeys the Second Law, that is the statement that entropy S is non-decreasing during any adiabatic process, not just quasi-static ones. A full axiomatization of the Second Law thus ought to include further axioms from which one can derive both Clausius' and Kelvin's statements of the Second Law. The former statement is

> *CLAUSIUS*: It is impossible in a cyclic process to transfer heat from a cold to a hotter reservoir, without any further changes taking place.

whereas the latter statement is

> *KELVIN*: It is impossible in a cyclic process to produce work by extracting heat from a single reservoir.

At first sight, the two statements are not quite similar. Clausius' statement concerns an asymmetry in heat exchange, in that heat transfers always proceed from hot to cold, i.e. it relates this asymmetry to the *ordering* of temperatures. Instead, Kelvin's statement makes explicit the idea that in order to produce work in a cycle one needs to exchange heat with at least two reservoirs at different temperatures. However, textbooks on thermodynamics standardly contain arguments that allegedly prove that the two statements of the Second Law are in fact logically equivalent. Afanassjewa (1925) was actually the first to recognize that this alleged equivalence presupposes that the absolute temperature T is assumed to be positive, and she added an *Axiom D*, or Temperature axiom, in order to capture this presupposition. That leads one to another question, though: that is, what happens when the absolute temperature becomes negative? We address this issue here in the final section.

The question of the emergence of the very concept of temperature (as distinct from the notion of heat), and the question of what would be an appropriate scale to express the concept numerically has an involved history, which goes back long before the rise of the theory of thermodynamics in the nineteenth century (see Chang 2004). The notion that one could, or indeed needs to distinguish between characterizing heat by an amount but also by an intensity, today called temperature, developed in the fifteenth and sixteenth century when the first thermometers were invented. But even when that idea took hold, the thermometers in actual use hardly led to a single universal scale because they relied on different thermometric substances (like mercury, alcohol, water, etc.), each of which had different properties in how they changed volume with temperature, and having different freezing and boiling points. Kelvin famously cut across all such considerations by proposing a scale for temperature that he called 'absolute', because it did not depend on the choice of any particular thermometric substance. In stead, he proposed to measure temperature by a scale that gave Carnot's theorem its most simple analytical expression. But even Kelvin did not suppose that 'absolute zero' was a barrier that could not be crossed.

This idea that absolute temperature has a lower bound of zero did become more persuasive, however, due to the concurrent rise in the 19th century of the kinetic theory, a microscopic theory with the purpose of underpinning thermodynamics, in which heat was seen as a 'kind of motion' and temperature was interpreted as the mean kinetic energy of the micro-constituents of a body. And since kinetic energy is by its very definition non-negative, the general consensus view emerged that absolute temperatures must be non-negative too in order to be physically meaningful. However, kinetic theory is only one special application of thermodynamics, and, as we will see below, there are other applications of thermodynamics where temperature is *not* correlated with mean kinetic energy. What is more, if we hold on to the view that thermodynamics is a macroscopic theory in its own right, not to be identified with what remains after the alleged reduction to kinetic theory has taken place, the question whether negative absolute temperatures are to be excluded as a matter of principle is not so obvious at all. Let us see how Afanassjewa's 1925 formulation of the Second Law of thermodynamics anticipated these outstanding issues.

She considered a Carnot process as a simple example of a cyclic process. (Recall that a Carnot cycle is composed of four quasi-static transformations: an isothermal one in which a quantity of heat Q_1 is absorbed from reservoir 1 at constant temperature T_1 and another in which heat Q_2 is emitted to a reservoir at lower constant temperature T_2, plus two intermediate transformations during which the gas is compressed and expanded adiabatically, thus without any exchange of heat.)

She first presented an argument, in footnotes 10 and 11 of her (1925), claiming that, if the temperature of the cold reservoir is set to zero, (i.e. $T_2 = 0$), this would provide a case in which Kelvin's formulation is violated, while Clausius' is not. This would then provide a first example in which these two formulations are not logically equivalent. However, this argument is complicated by the fact, mentioned in Sect. 3.2 above, that the absolute temperature T must be non-zero in order to act as an integrating divisor, a condition from which the existence of entropy is derived. If one allows states in Γ_{eq} for which $T = 0$, the entropy will become indeterminate in such states. And while Afanassjewa (1925) acknowledges this complication (in her footnote 10) the present authors believe the argument is not fully convincing.[11]

[11]Her argument conceives a Carnot cycle between two heat reservoirs, one with positive absolute temperature T_1, and the second with temperature $T_2 = 0$. In this cycle the system would absorb heat during the isothermal process in contact with the first reservoir, $Q_1 = T_1 \Delta S_1$, but would not exchange heat during the isothermal process with the second reservoir, since $Q_2 = T_2 \Delta S_2 = 0$. She argued that this would provide an example in which the Carnot cycle would provide work while only extracting heat from a single reservoir, in violation of Kelvin's statement of the Second Law. The problem with the argument is not just that that the definition of entropy by means of Pfaff's theorem fails when $T = 0$. The problem is also that some versions of the Third Law of Thermodynamics imply that states with absolute temperature $T = 0$ are physically unattainable, and thus could never be part of a Carnot cycle. Other versions of the Third Law imply that even if states of a system with $T = 0$ may be reached in a quasi-static process, their specific heat vanish at $T = 0$, making it impossible to exchange (non-zero) heat which such a system while keeping its temperature at $T = 0$. And although one can debate whether the Third Law (which Afanassewa (1925) does not discuss) is or is not part of classical thermodynamics, what is clear is that there are

However, her next example is much more intriguing and convincing. This example concerns a Carnot process, such that the temperature T_2 of the cold reservoir is *negative*, while T_1 is positive. (This avoids the complication associated with the integrating divisor issue which arises only when $T = 0$.) She argued that, in this case, the system actually absorbs heat Q_2 from the reservoir at temperature $T_2 < 0$ during the Carnot cycle, even if it emits entropy $\Delta S < 0$ to the reservoir, since $Q_2 = T_2 \Delta S > 0$. Of course, in this Carnot cycle, the total entropy change of the system is zero (in accordance with the requirement that $\oint dS = 0$) since the system also absorbs an entropy ΔS from the reservoir at temperature T_1 during the cycle. But the work W produced will be $Q_1 + Q_2$ rather than $Q_1 - Q_2$ for an ordinary Carnot process, and its efficency will be more than 100% since $\frac{W}{Q_1} = 1 - \frac{T_2}{T_1} > 1$.

She argued that this is in conflict with Clausius's statement.[12]

Afanassjewa hence claimed that while Axioms A, B and C are all necessary to derive the Clausius statement, this example showed they are not sufficient even taken together: one additional axiom is needed in order to rule out the possibility that heat flows from a colder to a hotter body. For this purpose, she proposed the following Temperature axiom:

> *Axiom D*. The absolute temperatures have one and the same sign for all states.

By requiring that when the higher temperature T_1 is positive the lower temperature T_2 must be positive too, one can complete the derivation of Clausius statement of the Second Law. Afanassjewa also noted that Axiom D is not needed to derive Kelvin's statement because the latter does not speak of the direction in which heat is transferred but only of the fact that more than one reservoirs be used. From this, she concluded: 'We therefore see that the two principles do not say the same thing and that one can derive neither Thomson's [i.e. Kelvin's] nor Clausius' principle from the existence of entropy alone' (p. 938).

It is a remarkable feat that Afanassjewa (1925) was able to show that Clausius and Kelvin statements are not logically equivalent when one admits both positive and negative absolute temperatures.[13] This, in turn, raises the further question whether

---

several ways in which one can deny the conclusion of her argument here (the violation of Kelvin's statement of the Second Law), while remaining consistent with her other axioms.

[12] Actually, she does not explain this conflict in detail. Perhaps the easiest way to construct a direct violation of the Clausius formulation of the Second Law is to combine the envisaged Carnot cycle with $T_2 < 0$ with another ordinary Carnot cycle, this second one between reservoirs at temperatures T_1 and T_3 , with $T_1 > T_3 > 0$ running in reverse. If one arranges this second cycle such that (i) it absorbs all the work produced in the first cycle, so that the combined cycle produces no work at all, and (ii) the amount of heat Q_1 that the first cycle absorbs from the heat reservoir at T_1 is balanced by the same amount of heat Q_1 being returned to that same reservoir during the second reversed Carnot cycle, the overall effect of the combined cycle will be one that produces no other change than to absorb heat Q_2 from the reservoir at temperature $T_1 < 0$ and emit the same amount of heat to the reservoir at temperature $T_3 > 0$. Thus, there would be a heat exchange in this combined cycle from a low temperature $T_2 < 0$ to a higher temperature $T_3 > 0$, without any further changes taking place.

[13] One may note that Carathéodory's (1909) axioms for thermodynamics actually also do not determine the sign of absolute temperature, but only upto arbitrary multiplicative and additive constants.

such formal conditions are physically possible, that is whether 'absolute zero' and 'absolute negative temperatures' can be achieved. Afanassjewa did not dwell too much into this issue. But she suggested that the question whether the conditions for negative absolute temperatures are physically possible can be answered by looking at the microscopic interpretation of thermodynamical systems. As she wrote in a footnote (numbered 14 in the translation):

> In the interpretation of absolute temperature in classical statistical mechanics, it naturally follows that this quantity can only be positive, in so far as it is the mean kinetic energy of molecules. However, whenever one is compelled —e.g. because of quantum theory— to deviate from this interpretation, Axiom D requires another special statistical interpretation. (p. 938)

The importance and originality of these considerations were borne out when 30 years later Ramsey (1956) pointed out concrete examples of quantum systems for which the absolute temperature can actually become negative.

The systems Ramsey considered are spin systems, whose Hamiltonian have a leading term that depends of the orientation of these spins in an external magnetic field. The system has a minimum value for its energy (obtained in the ground state when all spins are antiparallel to the magnetic field, but, —and this is the important point that distinguishes such systems from classical gases—it also has a maximum energy, in a 'ceiling state' where all spins are orientated parallel to the magnetic field. In a statistical treatment, roughly speaking, the entropy in both states is zero (because they can both be realized in a single way only). This means that if we plot the entropy as a function of energy, the entropy will be increasing with energy, until we reach the value halfway between the energy maximum and minimum, where the entropy attains it's maximum. (This is a state in which 50% of the spins are parallel and 50 % are antiparallel.) If we increase the energy further, the entropy starts to *decrease*, eventually dropping to zero in the ceiling state.

If one then assumes, as customary in statistical mechanics and thermodynamics, that the absolute temperature obeys the relation

$$\frac{1}{T} = \frac{dS}{dU} \tag{3.12}$$

one sees that the absolute temperature of the system will be negative on the decreasing flank of the entropy curve, growing from $-\infty$ at the midway point, to 0 at the ceiling state.

He argues that the choice of the multiplicative constant can be settled by convention, while the additive constant is to be determined "once and for all" by one single empirical observation of some irreversible process (p. 381), which together settle the sign of absolute temperature.

Although one might argue that Carathéodory thus also allowed for the conceivability of negative absolute temperatures, or at least pointed out that the sign of absolute temperature was not determined by his axioms alone, the fact that he considered this sign to be settled for all states and all systems by just one single empirical observation, shows that he did not conceive of the possibility of a system capable of having states with both positive and negative absolute temperature, let alone conceive of carrying out as Carnot cycle in such a system between them. In this respect, Carathéodory (1909) stands in marked contrast to Afanassjewa (1925).

In fact, Ramsey analyzed, just like Afanassjewa did 30 years earlier, Carnot processes between reservoirs of positive and negative temperature, and discussed the question whether the Kelvin and Clausius formulation of the Second Law are equivalent in this case. And just like Afanassjewa had done before him, he reached the conclusion that the answer is negative.

It is interesting to note, though, that while Afanassjewa argued the two formulations are inequivalent in this case in the sense that Clausius' formulation fails while Kelvin's formulation still holds; Ramsey argued for the opposite inequivalence, i.e. that Clausius' formulation holds while Kelvin's formulation fails. This distinction in their conclusions is mainly due to Ramsey's argument that, from a physical perspective one should regard thermal states of a spin system with negative absolute temperature as *hotter* than any thermal state with a positive temperature.[14] Thus, according to Ramsey, the ordering of temperature values in terms of the relation 'hotter than' or 'colder than' will not track the ordering of their numerical values. Rather, one would have to define that temperature T_1 is hotter than T_2 if and only if $-1/T_2 < -1/T_1$.

Ramsey's argument about the reordering of numerical temperature values is of course eminently reasonable for the kind of systems he considered. We do want to remark, however, that Afanassjewa obviously did not have such concrete examples at her disposal and was reasoning *in abstracto*, so that this idea of imposing a new convention in which the 'hotter than' ordering is separated from the numerical ordering could hardly have occurred to her. We also like to note that Ramsey's convention of temperature ordering skews the analysis a bit in Clausius' favour, because it actually *assumes* Clausius' claim that heat always flows from hotter to colder bodies. This claim is thereby turned from its original role as an empirical law into a defining condition of what it means for one temperature to be hotter than another and makes his claim (which is part of Clausius' formulation of the Second Law) true by definition.

However, in retrospect, it does not matter so much that Ramsey (1956) argued that Kelvin's formulation of the Second Law fails and Clausius' formulation holds, vis-a-vis a Carnot process with both positive and negative temperatures, while Afanassjewa argued for the opposite viewpoint; nor that she responded by including an axiom D that amounts to forbidding negative temperatures, where Ramsey displayed a concrete physical example of systems with negative temperatures. It remains striking that Afanassjewa analyzed the conceptual possibility of negative absolute temperatures and argued that such systems might be required in quantum theory and that admitting such states in thermodynamics would invalidate the equivalence between

[14]In simple terms, the argument is that if we would put two such systems into thermal contact, and allow them to equilibrate, this would bring about an equal distribution of energy. Since systems with negative temperature always have more energy per spin than systems with positive temperature, the heat flow in such an equilibration process will be from the system at negative temperature to the system of positive temperature. If one insists on Clausius' claim that heat always naturally flows from a hotter to a hotter than colder body, one concludes that a system with negative temperature will be hotter than any system with positive temperature.

Kelvin's and Clausius' formulation of the Second Law, some 30 years before this became an accepted view in the physics community. Once again, she appears to have been ahead of her times.

3.6 Conclusion

In this chapter we discussed Afanassjewa's work on the foundations of thermodynamics. We argued in the in Sects. 3.1 and 3.2 that her 1925 paper was inspired by the axiomatic approach to this theory by Carathéodory (1909) and was elaborated in a series of further papers (some of which remained unpublished) culminating in the 1956 publication of her book *Die Grundlagen der Thermodynamik*, a work that must have been largely composed in or before 1940.

Our discussion shows how Afanassjewa's work is particularly relevant to some themes in the current debate in philosophy of physics, on which we focused throughout the manuscript. The first theme hinges on her distinction between quasi-processes and quasi-static processes and we showed how this distinction can be employed to assuage an alleged paradox concerning thermodynamically reversible processes (Sect. 3.3). The second theme has to do with the distinction between heat and work, which she captured with an explicit axiom in her 1925 paper, being the first author to note this distinction and to elucidate it by means of an explicit example (Sect. 3.4). The third theme deals with the Second Law of Thermodynamics and the alleged logical equivalence between Clausius' and Kelvin's statements of this law: we have argued that, already in 1925, Afanassjewa noted that the equivalence holds only if one consistently adopts absolute temperatures with the same sign and that she was the first author to conceive of the possibility to perform a Carnot cycle on a single system between state of both positive and negative absolute temperatures, thereby anticipating a view that only became recognized in the physics community 30 years later (Sect. 3.5).[15]

Acknowledgements J.U. wants to express very special thanks to Tamara van Bommel, (Afanassjewa's granddaughter) for allowing me to browse through a box of Afanassjawa's papers and letters, and to Margriet van der Heijden for many helpful discussions and her assistance on this project.

[15] A referee referred us to two very relevant recent papers by David Lavis (2018, 2019). Lavis (2018) addresses Norton's paradox, which we discussed in Sect. 3.3 and discusses a proposal to solve it by what he calls the 'demarcation interpretation' of reversible processes, which seems very close to Afanassjewa's view in her (1956) book, but without referring to her work. The second paper, Lavis (2019) is most relevant to our discussion on negative temperatures in Sect. 3.5 and does refer to Afanassjewa's (1925) paper. And although Lavis (2019) agrees with her argument that Carnot cycles between two reservoirs at positive and negative temperature are possible, his paper disagrees with her conclusion that these would lead to a violation of Clausius' statement of the Second Law. The present authors believe, however, that this disagreement is due to the fact that Lavis adopts the Ramsey convention about how to order temperatures by the relation 'hotter than' (which makes any negative temperature hotter than any positive temperature), while Afanassjewa did not adopt this convention and regarded states at negative temperatures as colder than positive temperatures.

J.U. also thanks the Vossius Centre for the History of Humanities and Sciences at the University of Amsterdam and the Descartes Centre for the History and Philosophy of the Sciences and the Humanities at Utrecht University, the University of Geneva, the Polytechnical University of Milano, The University of Salzburg, and the Erwin Schrödinger Institute at the University of Vienna, for financial support and audiences at these various locations for their feedback.

References

Born, M. (1921). Kritische Betrachtungen zur traditionellen Darstellung der Thermodynamik. *Physikalische Zeitschrift*, *22*, 218–224, 249–254, 282–286.

Brown, H. R., & Uffink, J. (2001). The origins of time-asymmetry in thermodynamics: The minus first law. *Studies in History and Philosophy of Modern Physics*, *32*, 525–538.

Einstein, A. (1949). Autobiographical notes. In P. A. Schilpp (Ed.), *Albert Einstein, Philosopher-Scientist* (p. 33). Evanston Ill: The Library of Living Philosophers.

Carathéodory, C. (1909). Untersuchungen über die Grundlagen der Thermodynamik. *Mathematische Annalen*, *67*, 335–386.

Chang, H. (2004). *Inventing temperature: Measurement and scientific progress*, Oxford studies in the philosophy of science. New York: Oxford University Press.

Ehrenfest-Afanassjewa, T. (1925). Zur Axiomatisierung des Zweiten Hauptsatzes der Thermodynamik. *Zeitschrift für Physik*, *33*, 933–946.

Ehrenfest-Afanassjewa, T. (1936a). Over quasi-statische en niet-statische toestandsveranderingen. *Wis- en Natuurkundig Tijdschrift*, *8*, 29–34.

Ehrenfest-Afanassjewa, T. (1936b). Over omkeerbare en onomkeerbare processen. *Wis- en Natuurkundig Tijdschrift*, *8*, 38–40.

Ehrenfest-Afanassjewa, T. (1948). *Die Grundlagen der Thermodynamik; college van Mevr. Dr. T. Ehrenfest-Afanassjewa, 1947–1948* Universiteit Leiden.

Ehrenfest-Afanassjewa, T. (1956). *Grundlagen der Thermodynamik*. Brill: Leiden.

De Haas-Lorentz, G. L. (1938). *De beide hoofdwetten der thermodynamica en hare voornaamste toepassingen*. 's-Gravenhage: Nijhoff.

Lavis, D. A. (2018). The problem of equilibrium processes in thermodynamics. *Studies in History and Philosophy of Modern Physics*, *62*, 136–144.

Lavis, D. A. (2019). The question of negative temperatures in thermodynamics and statistical mechanics. *Studies in History and Philosophy of Modern Physics*. to appear.

Lieb, E. H., & Yngvason, J. (1999). The physics and mathematics of the second law of thermodynamics. *Physics Reports*, *310*, 1–96.

Norton, J. D. (2014). Finite idealizations. In *European philosophy of science–philosophy of science in Europe and the viennese heritage: Vienna circle institute yearbook* (Vol. 17, pp. 197–210). Springer: Dordrecht-Heidelberg-London-New York.

Norton, J. D. (2016). The impossible process: Thermodynamic reversibility. *Studies in History and Philosophy of Modern Physics*, *55*, 43–61.

Planck, M. (1926). Über die Begrundung des zweiten Hauptsatzes der Thermodynamik. *Sitzungsberichte der Preussischen Akademie der Wissenschaften, Physikalisch-Mathematische Klasse*, 453.

Ramsey, N. F. (1956). Thermodynamics and statistical mechanics at negative absolute temperatures. *Physical Review*, *103*, 20–28.

Perrin, J. B. (1913). *Les atomes*. Paris: Librairie Félix Alcan.

Thomsen, J. S., & Harka, T. J. (1962). Strange carnot cycles; Thermodynamics of a system with a density extremum. *American Journal of Physics*, *30*, 26–33.

Uffink, J. (2001). Bluff your way in the second law of thermodynamics. *Studies in History and Philosophy of Modern Physics*, *32*, 305–394.

Valente, G. (2019). On the paradox of reversible processes in thermodynamics. *Synthese*, *196*, 1761–1781.
Van der Waals, J. D., & Kohnstamm, Ph. (1927). *Lehrbuch der Thermostatik*. Leipzig: J.A. Barth.

Tatiana Afanassjewa and Paul Ehrenfest at their wedding in December 1904 in Vienna.
Source: M.J.Klein, Paul Ehrenfest; the Making of a theoretical Physicist, Amsterdam: Elsevier, 1970

The house that Afanassjewa designed, in a distinctly Russian style, in the Witterozenstraat 57, Leiden, in 1913. Characteristic of her design is that the house has a relatively closed facade, but opens up to the backside. View from the back. Source: Rijksdienst voor het Cultureel Erfgoed

The same house, seen from the streetside. In the facade there are two engravings. Source: J. Uffink

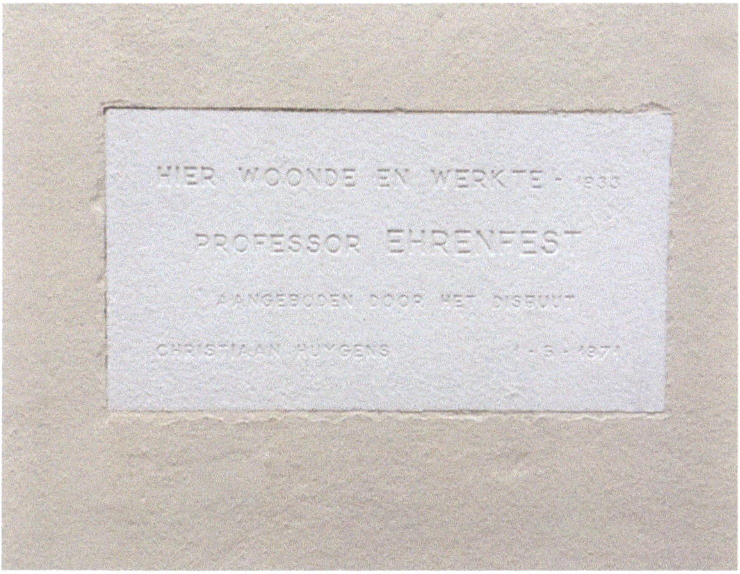

Two engravings on the facade of the house at Witterozenstraat 57. The translation of the first is "Here, Professor Ehrenfest lived and worked—1933. Offered by fraternity Christiaan Huygens, 1 May 1971"

The second engraving reads "His wife Tatiana Afanassjewa—far ahead of her times—opened this house for people and ideas". Source: J. Uffink

Tatiana Afanassjewa in front of Witterozenstraat 57 with daughter Galinka, around 1917. Source: family archive of Ehrenfest-Affanasjewa relatives

Afanassjewa teaching mathematics in Ordzhonikidze Ossetia, Soviet Union, 1933. Source: family archive of Ehrenfest-Affanasjewa relatives

Afanassjewa in her study in Leiden, 1956. Source: family archive of Ehrenfest-Affanasjewa relatives

A portrait of Tatiana Afanassjewa in 1954 by painter Harm Kamerlingh Onnes, a nephew of the Leiden physicist Heike Kamerlingh Onnes. Source: Museum de Lakenhal, Leiden

Part II
The Ehrenfests' Work on the Foundations of Statistical Mechanics

Chapter 4
Ehrenfest and Ehrenfest-Afanassjewa on Why Boltzmannian and Gibbsian Calculations Agree

Charlotte Werndl and Roman Frigg

4.1 Introduction

The relation between the Boltzmannian and the Gibbsian formulations of statistical mechanics (SM) is one of the major conceptual issues in the foundations of the discipline. In their celebrated review of SM, Paul Ehrenfest and Tatiana Ehrenfest-Afanassjewa discuss this issue and offer an argument for the conclusion that Boltzmannian equilibrium values agree with Gibbsian phase averages.[1] In this paper, we analyse their argument, which is still important today, and point out that its scope is limited to dilute gases.

[1]The original paper was published in German under the title 'Begriffiche Grundlagen der Statistischen Auffassung in der Mechanik' in 1911. Throughout this paper, we quote the English translation that came out in 1959 under the title 'The conceptual foundations of the statistical approach in mechanics'.

C. Werndl (✉)
Department of Philosophy (KGW), University of Salzburg, Franziskanergasse 1, 5020 Salzburg, Austria
e-mail: charlotte.werndl@sbg.ac.at

C. Werndl · R. Frigg
Department of Philosophy, Logic and Scientific Method, London School of Economics, Houghton Street, London WC2A 2AE, UK
e-mail: r.p.frigg@lse.ac.uk

J. Uffink et al. (eds.), *The Legacy of Tatjana Afanassjewa*, Women in the History of Philosophy and Sciences 7, https://doi.org/10.1007/978-3-030-47971-8_4

4.2 Boltzmannian and Gibbsian Statistical Mechanics

In statistical mechanics (SM) there are two main theoretical frameworks, namely Boltzmannian and Gibbsian SM.[2] Consider a system S consisting of the following: X is the set of all *possible states* (the state space), μ_X is the *probability measure* on X (that is assumed to be invariant under the dynamics) and $T_t(x)$ is the *dynamics* specifying the state of the system after t time steps given that it started in x.[3]

At the beginning of *Boltzmannian SM* stands the introduction of macro-states M_j, $j = 1, \ldots, m$, which are characterised by the values of a set of *macro-variables* $\{f_1, \ldots, f_k\}$ (where both m and k are in $\mathbb{N}$). A macro-variable $f_i : X \to \mathbb{R}$ is a function that associates a value with each $x \in X$. Capital letters F_i denote the values of the f_i. A *macro-state* M_i is defined by a particular set of values $\{F_1, \ldots, F_k\}$. Macro-states are assumed to supervene on micro-states, and hence there corresponds a micro-region $X_{M_j} \subseteq X$ to each M_j, which consists of all $x \in X$ for which the macroscopic variables assume the values characteristic for M_j. The X_{M_i} together form a partition of X, meaning that they do not overlap and jointly cover X. One of the macro-states is then singled out as the equilibrium state, and the *equilibrium values* of the f_i are the values F_i that the macro-variables assume in the equilibrium macro-state. The standard line on how to single out the equilibrium state is that size is the determining factor: the equilibrium state is the state for which $\mu_X(X_{M_i})$ assumes the highest value. As we will see in Sect. 4.5, this definition stands in need of qualification, but since it is widely used, we work with it for now and see how far it takes us.

The most important method to determine the largest macro-state is Boltzmann (1877) combinatorial argument, which Ehrenfest and Ehrenfest-Afanassjewa discuss in detail (1959, 26–30). The argument runs as follows. The state of one particle is given by a point in the six-dimensional state space X_1, and thus the state of the system (of the N particles) is given by N points in X_1. Because the system is confined to a finite container and the energy is constant, only a certain finite part of X_1 is accessible. This accessible part of X_1 is then divided into cells of equal size $\delta\omega$ whose dividing lines run parallel to the position and momentum axes. The result is a finite partition $\Omega := \{\omega_1, \ldots, \omega_l\}$, $l \in \mathbb{N}$. The cell in which a particle's state lies is referred to as the particle's coarse-grained micro-state. The specification of the coarse-grained micro-state for all particles is called an arrangement. Finally, a specification of the number of particles in each cell is referred to as a distribution $D = (N_1, N_2, \ldots, N_l)$ (N_i is the number of particles in cell ω_i). Each distribution is compatible with several arrangements, and the number of arrangements corresponding to a given distribution D is $G(D) = N! / N_1!N_2! \ldots, N_l!$.

[2] We briefly review both frameworks in this section. More extensive presentations can be found in Frigg (2008) and Uffink's (2007). See Frigg and Werndl (2019) for a discussion of the Gibbs formalism in particular.

[3] In this paper, we mostly follow Ehrenfest and Ehrenfest-Afanassjewa and consider deterministic systems. In our (2017) we discuss stochastic systems and show that the main results carry over to the stochastic context. We consider an explicitly stochastic system below in Sect. 4.6.

Ehrenfest and Ehrenfest-Afanassjewa now associate macro-states with distributions (1959, 49–50): each distribution defines a macro-state. This assumption is motivated by the fact that the macro-properties of a system are a function of the micro-properties, and hence a given macro-variable will assume different values for different distributions (we come back to this assumption below in Sect. 4.4). Clearly, every micro-state x of X corresponds to exactly one distribution $D(x)$. The macro-region X_D is then simply defined as the set of all x that are associated with the macro-state D.

The equilibrium macro-region is the region X_D with the largest measure. To determine this largest macro-region, Boltzmann (1877) provided a classical argument, which Ehrenfest and Ehrenfest-Afanassjewa discuss in detail (1959, 27–31). Boltzmann assumed that the energy e_i of particle i is only dependent on the cell in which it is located (and *not* on the location of the other particles), implying that the total energy of the system is $E = \sum_{i=1}^{l} N_i e_i$. With the further assumption that the number of cells in Ω is small compared to the number of particles, Boltzmann showed that $\mu_X(X_D)$ has a maximum when

$$N_i = \gamma e^{\lambda e_i}, \tag{4.1}$$

where γ and λ are parameters which depend on N and E. Equation (4.1) is now known as the *discrete Maxwell–Boltzmann distribution*. The equilibrium macro-state, therefore, corresponds to the Maxwell–Boltzmann distribution.

However, as Ehrenfest and Ehrenfest-Afanassjewa rightly emphasise (1959, 30), there is a last step missing. The X_D as defined above are $6N$-dimensional, and Eq. (4.1) gives us is the distribution for the cell of largest size relative to the Lebesgue measure μ_X (or more precisely, relative to the $6N$-dimensional subset X_{ES} of X defined by the condition that $E = \sum_{i=1}^{l} N_i e_i$). However, by assumption, the system has constant energy, and so we know that the system's motion takes place on the $6N$-1-dimensional energy hypersurface X_E. Hence, the relevant macro-regions are ones that lie in X_E rather than in X. A quick fix is the following: define the relevant $6N$-1-dimensional macro-regions as the intersection of the $6N$-dimensional X_D with X_E, and use the restriction μ_E, the restriction of μ_X to X_E, to measure their size.

Ehrenfest and Ehrenfest-Afanassjewa are careful to point out that this is not enough to give us what is needed, namely the macro-region of largest size relative to the measure μ_{X_E} on the $6N-1$-dimensional set X_E. Standard presentations of the combinatorial argument simply assume that the possible distributions and the proportion of the different distributions would not change if macro-states were instead defined on X_E, which yields the desired result that the equilibrium region is the largest region on X_E. Ehrenfest and Ehrenfest-Afanassjewa (1959, 30) are more careful. While they also adopt this assumption, they stress that it is in need of further justification.

So the conclusion Ehrenfest and Ehrenfest-Afanassjewa arrive at is that *in the Boltzmannnian framework the observed value in equilibrium for the observable f is the value of f in the macro-region corresponding to the Maxwell–Boltzmann distribution.*

Gibbsian SM studies *ensembles*, infinite collections of independent systems that are all governed by the same equations but start in different initial states. Formally, an ensemble is a probability density $\rho(x, t)$, $x \in X$, describing the probability of finding the state of a system chosen at random from the ensemble in a certain region of X at time t.

Given an ensemble ρ, the Gibbs entropy is

$$S_G[\rho] = -k_B \int_X \rho(x, t) \log[\rho(x, t)] \mathrm{d}x, \tag{4.2}$$

where k_B is the Boltzmann constant. An ensemble $\rho(x, t)$ is called *stationary* if and only if it does not depend on time, i.e. $\rho(x, t) = \rho(x)$ for all t. In Gibbsian SM equilibrium is a property of an ensemble. More specifically, the ensemble is in equilibrium if and only if it is stationary, and sometimes it is also required that it has maximum Gibbs entropy given the constraints imposed on the system. The most common constraints give rise to the microcanonical, canonical and grand-canonical distributions (1959, 46–47).

As in Boltzmannian SM, physical observables correspond to a set of real-valued functions f_i, and the *phase average* of such a function in equilibrium is defined as

$$\langle f_i \rangle = \int_X f_i(x) \rho(x) \mathrm{d}x. \tag{4.3}$$

According to the canonical understanding of Gibbsian SM, *what is observed in experiments on systems in equilibrium are such phase averages* (1959, 47 and 49). There is, however, a question about the scope of this claim: according to Gibbsian SM, does one *always* observe phase averages or are phase averages only observed in certain situations? The answer to this question is a matter of dispute which depends on how exactly Gibbsian SM is interpreted (for a discussion see Frigg and Werndl 2019). It is not entirely clear what reading of Gibbsian SM Ehrenfest and Ehrenfest-Afanassjewa endorse (though it seems to us that they rather endorse the claim and that, according to Gibbsian SM, always phase averages are observed). Fortunately, this issue does not matter in what follows.

Now, we are in a curious situation. Two different frameworks make predictions for the same experimental values. The Boltzmannian account says that the observed equilibrium value for the observable f_i is the value that it assumes in the macro-region corresponding to the Maxwell–Boltzmann distribution, while the Gibbsian account says that that the equilibrium value is $\langle f_i \rangle$. Do these values coincide? If so, why? If not, which of the values, if any, is correct?

4.3 Ehrenfest and Ehrenfest-Afanassjewa on Gibbs Versus Boltzmann

Ehrenfest and Ehrenfest-Afanassjewa opt for the first solution and set out to show that Boltzmannian equilibrium values and Gibbsian phase averages coincide. Their argument is an important one, and similar points have been made more recently by Davey (2009), Myrvold (2016). They begin by discussing the Gibbsian treatment of the gas with the observable f.[4] According to the Gibbsian framework, what is observed in equilibrium is the phase average. Because energy is conserved, it would be natural to consider the phase average relative to the micro-canonical ensemble (because this is the stationary distribution of maximum Gibbsian entropy under the constraint of constant energy). However, Ehrenfest and Ehrenfest-Afanassjewa do not do this and instead consider the phase average with respect to the canonical ensemble. The canonical ensemble is the stationary distribution of maximum entropy when the energy is allowed to vary:

$$\rho_c(q, p) = e^{\frac{\Psi - E(q,p)}{\Theta}}, \tag{4.4}$$

where $E(q, p)$ is the total energy, Θ is an constant, and Ψ is determined by the constraint that $\int_X \rho_c(q, p) = 1$.

The reason why they consider the phase average with respect to the canonical ensemble is unclear. A possible motivation might be that they want to show that it does not matter which distribution is chosen: Gibbsian SM leads to the same result as Boltzmannian SM regardless of whether one works with the microcanonical or the canonical ensemble.

As a first step they appeal to the well-known result, often referred to as the equivalence between the microcanonical and canonical distributions that holds when the number of particles of a gas is extremely large:

> In an ensemble which is canonically distributed with the modulus $\Theta = \Theta_0$, an overwhelming majority of individuals will have nearly the same total energy $E = E_0$ (Ehrenfest and Ehrenfest-Afanassjewa 1959, 48).

(Here Θ_0 is the fixed value of Θ in Eq. (4.4) of the canonical distribution above and E_0 is the energy value that nearly all individuals will have for the fixed value Θ_0).

Based on this result Ehrenfest and Ehrenfest-Afanassjewa (1959, 48–49) argue that it is plausible that $\int_X f(x) d\rho_c$, the phase average with respect to the canonical distribution ρ_c on X, is approximately equal to $\int_{X_E} f(x) d\rho_m$, the phase average with respect to the microcanonical distribution ρ_m on X_E (when f is restricted to X_E).

The next step is the vital move in the argument. Recall that the combinatorial argument shows that the equilibrium macro-region is the largest macro-region. So the macro-value corresponding to the Maxwell–Boltzmann distribution is the macro-value that is taken by more microstates than any other macro-value on X_E.[5] It is

[4] For ease of notation, we suppress the subscript 'i' from now.

[5] Strictly speaking, this is true only under an additional assumption that we discuss in the next section.

crucial to be clear on the sense of 'large' that is being used here. What the combinatorial argument shows is that the equilibrium macro-region is larger than *any other macro-region*. It does not show that the equilibrium macro-region is large in an absolute sense, i.e. that it occupies the largest part of X_E. The latter does not follow from the former. A macro-region can be larger than any other macro-region without being large relative to X_E. Ehrenfest and Ehrenfest-Afanassjewa bridge the gap between a relative and the absolute sense of 'large' by referring to results due to Jeans (1904, Sects. 46–56), who argues that *nearly all* states in X_E are in the macro-region corresponding to the Maxwell–Boltzmann distribution. Hence, f assumes the equilibrium value on almost all states in X_E. From this, they infer that this value is approximately equal to the Gibbsian phase average derived in the previous paragraph.

So their conclusion is that in a system in which the combinatorial argument applies, the Boltzmannian equilibrium value and the Gibbsian phase average with respect to the macro-variable f are approximately the same.

4.4 Assessment of Ehrenfest and Ehrenfest-Afanassjewa's Argument

The considerations we make to assess the Ehrenfest and Ehrenfest-Affanassjewa argument fall into two groups. Considerations in the first group concern the combinatorial argument and its limitations; considerations in the second group concern the identity argument in the last section. We will focus mainly on the second group but will begin by making a few observations about the first.

As has been pointed out previously,[6] a core assumption of the combinatorial argument, namely that $E = \sum_{i=1}^{l} N_i e_i$, is very restrictive. In essence, this assumption implies that the argument only applies (even in an approximate form) to *dilute gases*. So it is unsurprising that Ehrenfest and Ehrenfest-Afanassjewa (1911, 36–60) talk about gas systems when presenting the combinatorial argument. However, it remains unclear from the text whether they are clear on the fact that it *only* applies to dilute gases.

Second, the conclusion that the macro-value of f in the Maxwell–Boltzmann distribution is the macro-value that is taken by more micro-states than any other macro-value on X_E follows only under the strong assumption that f assumes a different value for *every* macro-region. However, Lavis (2005, 2008) pointed out that this need not always be the case.[7] Macro-regions can show *degeneracy* in the sense that f can assume the same value in several regions. It is possible that a number of such (non-equilibrium) macro-regions *taken together* are larger than the equilibrium region, and so f assumes the equilibrium value in a region of the state space that is smaller than the union of the degenerate macro-regions. Lavis (2005,

[6]See, for instance, Uffink (2007) and Werndl and Frigg (2015b).

[7]Lavis (2005, 2008) discussed the case of the Boltzmann entropy, but the point obviously generalises to phase functions.

2008) shows that this happens in the case of the baker's gas, thereby driving home the point that degeneracies causing difficulties is more than just a theoretical possibility.

Let us set these concerns aside and assume, for the sake of argument, that we are dealing with a dilute gas and a 'well-behaved' function f (we will discuss what happens if these assumptions fail in Sect. 4.5). Does Ehrenfest and Ehrenfest-Afanassjewa's equivalence argument hold under these assumptions? It is obvious that their argument contains a gap. They conclude from the fact that f assumes the equilibrium value on nearly all states in X_E that the average of f over X_E is approximately equivalent to that value. This, however, is true only if the non-equilibrium values are not disproportionately far away from the equilibrium value. If the non-equilibrium values differ significantly from the equilibrium values, their contribution to the average can be significant and the average need no longer be equal to the equilibrium value of the function, not even approximately.

To rule out such a scenario one needs to assume that f satisfies some kind of 'small fluctuation condition'. The most common condition of this kind is now known as the *Khinchin Condition.* The condition plays a crucial role in the work of Khinchin (1960 [1949]) and variants of it have been appealed to in the foundational literature on SM, for instance by Malament and Zabell (1980), Myrvold (2016). This condition requires that the observable f equals the phase average nearly everywhere on phase space. Formally:

> There is a $\bar{X} \subseteq X$ with $\mu_X(\bar{X}) = 1 - \delta$ for a small $\delta \geq 0$ such that $|f(x) - \langle f(x) \rangle| \leq \varepsilon$ for all $x \in \bar{X}$ and a very small $\varepsilon \geq 0$.

Under Ehrenfest and Ehrenfest-Afanassjewa's assumptions the Boltzmannian equilibrium macro-region satisfies the condition on $\bar{X}$. Let F_{equ} be the value of f in that macro-region. It then follows that $|\langle f(x) \rangle - F_{equ}| \leq \varepsilon$, and therefore the Boltzmannian value and the Gibbsian average agree, at least approximately.

Ehrenfest and Ehrenfest-Afanassjewa, however, do not appeal to this formulation of the condition, but to a variant of the Khinchin condition that we call the *Ehrenfest-Afanassjewa Condition*. The condition is that the observable f is approximately equal to the Boltzmannian equilibrium value nearly everywhere on phase space and that the observable does not take extreme values on the rest of the phase space. Formally, the Ehrenfest-Afanassjewa Condition can be formulated as follows[8]:

> Consider a system of the kind introduced in Sect. 4.2 endowed with an observable f. Further assume that the system has a Boltzmannian equilibrium with equilibrium macro-value F_{equ}.

---

[8]A variant of the Ehrenfest-Afanassjewa Condition requires that the observable f is constant nearly everywhere on phase space and does not take extreme values on the rest of the phase space:

> There is an constant $C \in \mathbb{R}$ and a $\bar{X} \subseteq X$ with $\mu_X(\bar{X}) = 1 - \delta$ (for a small $\delta \geq 0$) such that (i) $|f(x) - C| \leq \varepsilon$ for all $x \in \bar{X}$ for a very small $\varepsilon \geq 0$ and (ii) $|\int_{X \setminus \bar{X}} f(x) d\mu_X - C\delta)| \leq \gamma$ (for a very small $\gamma \geq 0$).

Because the Boltzmannian equilibrium macro-value F_{equ} takes up more than δ of phase space, it follows that F_{equ} is very close to C. Therefore, $|f(x) - F_{equ}| \leq \epsilon_1$ for a small $\varepsilon_1 \geq 0$ for all $x \in \bar{X}$ and $|\int_{X \setminus \bar{X}} f(x) d\mu_X - F_{equ}\delta)| \leq \gamma_1$ (for a very small $\gamma_1 \geq 0$). This is in fact the original Ehrenfest-Afanassjewa Condition and so the variant is in fact equivalent to the original Ehrenfest-Afanassjewa Condition.

Then there is an $\bar{X} \subseteq X$ with $\mu_X(\bar{X}) = 1 - \delta$ (for a small $\delta \geq 0$) such that (i) $|f(x) - F_{equ}| \leq \varepsilon$ for all $x \in \bar{X}$ (for a very small $\varepsilon \geq 0$) and (ii) $|\int_{X \setminus \bar{X}} f(x) d\mu_X - F_{equ}\delta)| \leq \gamma$ (for a very small $\gamma \geq 0$).

A simple calculation shows that for systems that satisfy the Ehrenfest-Afanassjewa Condition with respect to f, the phase average is approximately equal to the Boltzmannian equilibrium macro-value F_{equ}:

$$|\langle f(x) \rangle - F_{equ}| \leq$$

$$|\int_{\bar{X}} f(x) d\mu_X - F_{equ}(1 - \delta)| + |\int_{X \setminus \bar{X}} f(x) d\mu_X - F_{equ}\delta| \leq$$

$$\varepsilon(1 - \delta) + \gamma \text{ (because of (i) and (ii) of the Khinchin condition).}$$

It is interesting to discuss both the Khinchin and the Ehrenfest-Afanassjewa conditions because, depending on the context, one or the other may turn out to be more useful. There is, however, a slight mismatch between the Ehrenfest-Afanassjewa Condition and the calculations of Ehrenfest and Ehrenfest-Afanassjewa: they perform Gibbsian phase space averaging with the canonical and not the micro-canonical distribution. However, because of the equivalence of the micro-canonical and macro-canonical ensemble as discussed above this difference does not matter; and if for some reason it did, one could simply perform the Gibbsian calculations with the microcanonical ensemble.

It is important to note that neither of the two conditions is in any way trivially true. Khinchin could prove his condition only for the special case of sum functions in non-interacting systems (sum functions are functions in many-particle systems that can be written as a sum over one-particle functions). The generalisation of this result to the case interacting system is a veritable challenge and no general solution has been found to date.[9]

Ehrenfest and Ehrenfest-Afanassjewa argue in their survey that the Ehrenfest-Afanassjewa condition is satisfied. Their argument is valid but only subject to a change in one of the assumptions and an additional assumption in their argument. Namely, first, as outlined above, they assume (by referring to Jeans 1904, Sects. 46–56) that nearly all states in X_E are in the macro-region corresponding to the Maxwell–Boltzmann distribution. We have seen above that this need not always be the case. Furthermore, a closer look at Jeans' text reveals that he does not actually offer a proof of the claim. What Jeans shows is that the nearly all of phase space X is taken up by macro-regions with a distribution D *very close* to the Maxwell–Boltzmann distribution. Hence the assumption that the macro-region corresponding to the exact Maxwell–Boltzmann distribution is large in absolute terms has to be given up. Fortunately, a weaker assumption provides what we need. All that is required for the argument to go through is that the observable f is such that macro-regions with distribution D very close to the Maxwell–Boltzmann distribution have approximately the

[9]See Uffink's (2007, 1020–1028) for a discussion.

same macro-value as the macro-regions with the Maxwell–Boltzmann distribution.[10] Note that this amounts to conditions imposed on the Boltzmannian macro-structure f.

With this new assumption in place, Jeans' (1904, Sects. 46–56) calculations indeed imply that condition (i) of the Ehrenfest-Afanassjewa Condition is satisfied. Second, Jeans (1904, Sects. 46–56) shows that the states whose macro-values are not very close to the Maxwell–Boltzmann distribution take up a tiny fraction of phase space, i.e. $X \setminus \bar{X}$ is extremely small. But what is still needed is the further condition that f does not take extremely large or extremely low values on $X \setminus \bar{X}$ (and this again is a condition imposed on f). With this new assumption in place, (ii) of the Ehrenfest-Afanassjewa condition is satisfied. Hence, we conclude that *with the modifications just outlined the Ehrenfest-Afanassjewa condition is satisfied and the Boltzmannian equilibrium value and the Gibbsian phase average lead to approximately the same result*.

To sum up, Ehrenfest and Ehrenfest-Afanassjewa identify an important case where the Boltzmannian equilibrium values and the Gibbsian phase averages agree. However, their argument relies on strong assumptions, and while these assumptions are satisfied for certain observables in the case of dilute gases, the assumptions need not hold in general. In fact, in the remainder of this paper, we discuss cases that do not fit Ehrenfest and Ehrenfest-Afanassjewa's mould. First, there are cases where the Boltzmannian equilibrium value is different from the Gibbsian phase average. This shows that it is an important task for foundational debates to find out under what conditions the Boltzmannian equilibrium value and the Gibbsian phase average agree or disagree. Examples of disagreement will be discussed in Sect. 6. Second, there are cases where the Boltzmannian equilibrium value and the Gibbsian phase averages agree but where the Ehrenfest-Afanassjewa condition is not satisfied. The Ehrenfest-Afanassjewa condition and the Khinchin condition provide one condition where there is an agreement (cf. also Werndl and Frigg 2017a; 2017b, 2020).

For instance, consider the Kac ring, consisting of an even number N of sites distributed equidistantly around a circle. On each site, there is a spin, which can be in states up (u) or down (d). A *micro-state* x^{kr} of the Kac ring is a specific combination of up and down spin for all sites and the full state space $Z = K^{kr}$ consist of all combinations of up and down spins (i.e. of 2^N elements). There are s, $1 \leq s \leq N-1$, spin flippers distributed at some of the midpoints between the spins. The dynamics rotates the spins one spin site in the clockwise direction every second (or whichever unit of time one chooses), and when the spins pass through a spin flipper, they change their direction. The measure that is usually considered is the uniform measure $\mu_{X^{kr}}$ on X^{kr} (Lavis 2008). The *macro-states* usually considered are the *total number of up spins*, conveniently labelled as M_i^K, where i denotes the total number of up spins, $0 \leq i \leq N$. The Kac-ring with the standard macro-state structure is a paradigm example where Boltzmannian equilibrium values and Gibbsian phase averages agree.

[10] Given a certain macro-variable f and an allowable difference between the Gibbsian phase average and the Boltzmannian equilibrium macro-value, one could precisely quantify what notion of 'approximately the same macro-value as the Maxwell–Boltzmann distribution' would be needed in order for the Khinchin theorem to go through by making use of the calculations in Jeans (1904, Sects. 46–56).

However, it is not an instance of the Ehrenfest-Afanassjewa-condition because, as shown in Lavis (2005, 2008), the equilibrium macro-region corresponding to an equal number of up and down spins only takes up less than half of state space (the rest is taken up by macro-states that are macroscopically distinguishable from the Boltzmannan equilibrium macro-state). Other examples where the Boltzmannian equilibrium value and the Gibbsian phase average agree but where the Ehrenfest-Afanassjewa condition the Khinchin condition does not apply include the baker's gas with the standard macro-state structure and the ideal gas with the standard macro-state structure (cf. Werndl and Frigg 2017a; 2017b, 2020). The reason why the Boltzmannian equilibrium value and the Gbbsian phase average agree in these cases will be discussed later in Sect. 4.7.

4.5 Beyond Dilute Gases

As we have seen above, the combinatorial argument is restricted to dilute gases. Most systems of interest in SM are not of this kind and so this is a serious restriction. In two recent papers, we have discussed this problem at length and proposed an alternative Boltzmannian definition of equilibrium (2015a, 2015b). On this definition, it is not size but 'residence time' that defines equilibrium: the macro-state in which the system spends most of its time is the equilibrium macro-state. More specifically, define LF_R to be the fraction of time a system spends in region $R \subseteq X$ in the long run:

$$LF_R(x) = \lim_{t\to\infty} \frac{1}{t} \int_0^t 1_A(T_\tau(x))\mathrm{d}\tau, \tag{4.5}$$

where $1_A(x)$ is the characteristic function of R: $1_A(x) = 1$ for $x \in R$ and 0 otherwise.

'Most' is interpreted as requiring that the system spends more time in equilibrium than in any other macro-state, leading to the notion of an γ-ε-equilibrium[11]:

> Let $\gamma > 0$ and let ε be a very small positive real number, $\varepsilon < \gamma$. If there is a macro-state $M_{F_1^*,\ldots,F_l^*}$ satisfying the following condition, then it is the γ-ε-equilibrium state of S: There exists a set $Y \subseteq X$ such that $\mu_X(Y) \geq 1 - \varepsilon$, and all initial states $x \in Y$ satisfy $LF_{X_{M_{F_1^*,\ldots,F_l^*}}}(x) \geq LF_{X_{M_{F_1,\ldots,F_l}}}(x) + \gamma$ for all macro-states $M \neq M_{F_1^*,\ldots,F_l^*}$

Clearly, *the value observed in equilibrium is simply the value associated with the equilibrium macro-state*. Further, it should be mentioned that one can prove that equilibrium states defined in this way correspond to the largest macro-region in the sense that their measure is $\gamma - \epsilon$ larger than any other macro-region (Werndl and

[11] Alternatively, 'most' can also be understood as referring to the fact that the system spends at least $\alpha > 1/2$ of its time in equilibrium, leading to the different notion of an α-ε-equilibrium. Nothing in what follows hinges on which notion of equilibrium is adopted (cf. Werndl and Frigg 2015b and forthcoming references).

Frigg 2015b). This provides a notion of equilibrium that is fully general in that it does not depend on the system's dynamics and is hence applicable also to strongly interacting systems like solids and fluids.

4.6 An Example Where Boltzmannian Equilibrium Values and Gibbsian Phase Averages Differ

In this section, we see that Ehrenfest and Ehrenfest-Afanassjewa's result fails to generalise: in strongly interacting systems like solids and fluids the Boltzmannian equilibrium value and the Gibbsian phase average can differ. The six-vertex model with energy as the relevant macro-variable will serve as an example of a case where the Boltzmannian and Gibbsian equilibrium values differ. Consider a two-dimensional quadratic lattice with N sites on a torus (the choice of a torus ensures that every grid point has exactly four nearest neighbours, thus allowing to neglect border effects). Each site is connected to its four nearest neighbours by edges. Each edge carries an arrow that either points towards or away from the site. The so-called 'ice-rule' restricts the allowable arrangements of the arrows: the arrows have to be distributed in a way such that at each site in the lattice there are exactly two inward and two outward pointing arrows. It is easy to see that there are exactly six configurations of the arrows that satisfy the ice-rule, and they are shown in Fig. 4.1. The name 'six-vertex model' is motivated by the existence of these six configurations.

The reason for the name 'ice-rule' is that in frozen water each oxygen atom is connected to four other oxygen atoms. So the sites can be thought of as representing oxygen atoms and the edges as representing their bonds. For each bond, there is a hydrogen atom that does not sit in the middle between the two oxygen atoms but instead occupies a position closer to one of the oxygen atoms. Thus, the arrows can be interpreted as indicating to which oxygen atom the hydrogen atom is closer. The ice-rule then corresponds to the requirement that each oxygen atom has two close and two remote hydrogen atoms. Not only water ice but also several crystals, in particular potassium dihydrogen phosphate, satisfy the ice-rule (cf. Baxter 1982; Lavis and Bell 1999; Slater 1941).

The micro-states of the six-vertex model $\xi = (\xi_1, \ldots, \xi_N)$ are given by assigning one of the six types of configurations of the arrows permitted by the ice rule to each site in the model. Each of the six configurations has a certain energy ϵ_j, $1 \leq j \leq 6$.

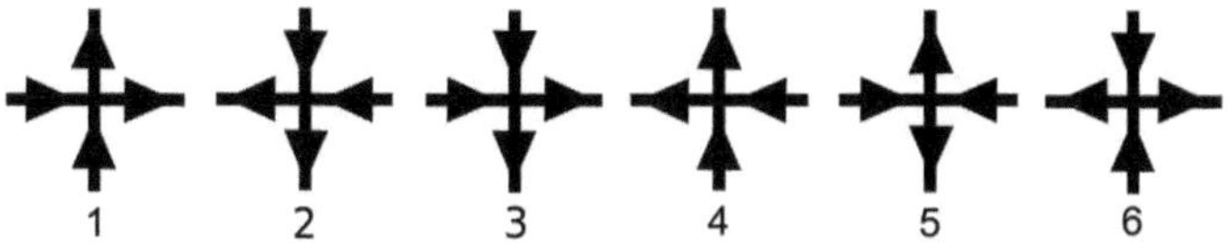

Fig. 4.1 The configurations of the six-vertex model

Denote by $\epsilon(\xi_j)$ the energy of the jth configuration. Then the energy of the state ξ is given by:

$$E(\xi) = \sum_{i=1}^{N} \epsilon(\xi_i). \tag{4.6}$$

We now assume that the energy of the different configurations is $\epsilon_1 = \epsilon_2 = 0$ and $\epsilon_3 = \epsilon_4 = \epsilon_5 = \epsilon_6 = 1$. The probability of the micro-states is given by the canonical distribution $p(\xi) = e^{-E(\xi)/kT}/Z$, with $Z = \sum_{\xi} e^{-E(\xi)/kT}$. Note that this is merely the probability measure over the micro-states, and is per se neither Boltzmannian nor Gibbsian. For the six-vertex model, one usually works with a stochastic dynamics. More specifically, the underlying dynamics is assumed to be an irreducible Markov chain (Baxter 1982; Lavis and Bell 1999; Werndl and Frigg 2020). The probability $p(\xi)$ is then invariant under the Markov dynamics and is thus a stationary probability measure.

We now study the six-vertex model with the *internal energy* E as defined in Eq. (4.6) as the relevant macro-variable for low temperatures. The lowest energy value is $E = 0$, which defines a macro-state M_0 with macro-region $X_{M_0} = \{\xi^*, \xi^+\}$ (here ξ^* is the state where all vertices are in the first configuration, and ξ^+ is the state where all vertices are in the second configuration). Note that the lower the temperature, the larger the probability of the lower energy states; and the higher the temperature, the more uniform the probability measure. Hence for sufficiently low temperatures, the probability mass is concentrated on low-energy states. For this reason, X_{M_0} is the largest macro-region. Because the dynamics is an irreducible Markov chain, the model spends most of its time in M_0. It follows that M_0 is the Boltzmannian equilibrium state and $E = 0$ is the Boltzmannian equilibrium value (cf. Werndl and Frigg 2020).

Let us now turn to the Gibbsian treatment. Here, $p(\xi)$ is the stationary measure of maximum entropy, and E is observable. E will assume its lowest value $E = 0$ only for two specific micro-states, namely ξ^* and ξ^+. For all other states (and they all have positive probability), the value of E will be higher. From this, we conclude that the Gibbsian phase average $\langle E \rangle$ is greater than zero and hence higher than the Boltzmannian equilibrium value. Thus, the Boltzmannian equilibrium value and the Gibbsian phase average differ.

Now, of course, the question is whether this difference can be significant. To see that this can be so, choose a T such that $\{\xi^*, \xi^+\}$ is still the largest macro-region but that the probability of this macro-region is equal or less than 0.5.[12] Clearly, the Boltzmannian equilibrium value is still $E = 0$. Yet the second lowest macro-value is $E = \sqrt{N}$, which is the energy corresponding to micro-states where all columns of the lattice except one are taken up by states which are in the first or the second configuration, and the states in the exceptional row are all states in the third or

[12] As we have seen, for sufficiently low temperatures $\{\xi^*, \xi^+\}$ is the largest macro-region. The higher the temperature, the more uniform is the probability measure. Hence, for sufficiently high temperatures, the largest macro-region will differ from $\{\xi^*, \xi^+\}$. Because the canonical distribution is continuous in T, there exists a T such that $\{\xi^*, \xi^+\}$ is the largest macro-region but its probability is ≤ 0.5.

fourth configuration.[13] It follows that $\langle E \rangle$ is higher than $\sqrt{N}/2$. Consequently, the Gibbsian phase average and the Boltzmannian equilibrium value will differ by more than $\sqrt{N}/2$, which is not a difference that is negligible (especially when N is large). Note also that the Boltzmannian macro-value that is closest to the value obtained from Gibbsian phase space averaging is larger or equal to $\sqrt{N}$. But this Boltzmannian macro-value is *different* from the Boltzmannian equilibrium macro-value, which is zero. This again underlines that Gibbsian phase space averaging results in a different outcome than the Boltzmannian calculations.[14]

4.7 When Boltzmann and Gibbs Agree

Boltzmannian equilibrium values and Gibbsian phase averages can come apart. This raises the question under what conditions the two coincide. We have already seen in Sect. 4.4 that one situation where there is agreement is when the Ehrenfest-Afanassjewa condition is satisfied. However, as already noted then, there are important cases including the baker's gas with the standard macro-state structure, the KAC-ring with the standard macro-state structure and the ideal gas with the standard macro-state structure, that do not, in general, satisfy the Ehrenfest-Afanassjewa condition or the Khinchin condition.

In our (2017b, 2020) we present *another* set of conditions under which the Boltzmannian equilibrium value and the Gibbsian phase average coincide. Intuitively speaking, the conditions are: (i) the measure on phase space is the product measure of the one-constituent space; (ii) the macro-variable considered is the sum of a one-constituent observable; and (iii) this one-constituent observable takes finitely many values with the same probability. With these conditions in place, the average equivalence theorem then shows that, if a Boltzmannian equilibrium exists, the Boltzmannian equilibrium value and the Gibbsian phase average coincide:

Average Equivalence Theorem (AET). Suppose that a system with phase space X, dynamics T_t and measure μ_X is composed of $N \geq 1$ constituents. That is, the state $x \in X$ is given by the N coordinates $x = (x_1, \ldots, x_N)$; $X = X_1 \times X_2 \ldots \times X_N$, where $X_i = X_{oc}$ for all i, $1 \leq i \leq N$ (X_{oc} is the one-constituent space). Let μ_X be the product measure $\mu_{X_1} \times \mu_{X_2} \ldots \times \mu_{X_N}$, where $\mu_{X_i} = \mu_{X_{oc}}$ is the measure on X_{oc}. Suppose that an observable κ is defined on the one-particle space X_{oc} and takes the values $\kappa_1, \ldots, \kappa_k$ with equal probability $1/k$, $k \leq N$.[15] Suppose that the macro-variable K is the sum of the one-component observable, i.e. $K(x) = \sum_{i=1}^{N} \kappa(x_i)$. Then the value corresponding to the largest macro-region as well as the value obtained by phase space averaging is $\frac{N}{k}(\kappa_1 + \kappa_2 + \ldots \kappa_N)$.

---

[13]Such micro-state corresponds to the smallest possible departure from the macro-state with zero energy because the number of downward pointing arrows is the same for all rows. From this, then follows that there has to be a perturbation in each row and that $\sqrt{N}$ has to be the second lowest value of the internal energy (Lavis and Bell 1999).

[14]Further examples where the Gibbsian phase average and the Boltzmannian equilibrium value come apart can be found in our (2017b and 2020).

[15]It is assumed that N ia a multiple of k, i.e. $N = k * s$ for some $s \in \mathbb{N}$.

This theorem applies to the KAC-ring and the other examples (baker's system, ideal gas) mentioned above as cases where the Boltzmannian equilibrium value agrees with the Gibbsian phase average but where the Ehrenfest-Afanassjewa condition does not apply. Hence, it explains in these cases why the Boltzmannian equilibrium value and the Gibbsian phase average coincide. As it should be, the theorem does not apply to the six vertex model with the energy macro-variable because conditions (i) and (iii) are not satisfied (the measure is not the product measure of the one-constituent space, and the macro-variable considered is not the sum of a one-constituent observable, taking values with equal probability).

Note that the conditions of the Average Equivalence Theorem are not necessary for Boltzmannian equilibrium values and Gibbsian phase averages to coincide. In particular, that the macro-variable is a sum of the variables on the one-component space, that the macro-variable on the one-component space corresponds to a partition into cells of equal probability, or that the measure on state space is the product measure of the measure on the one-component space are strong conditions that are often not satisfied. This is illustrated by our example of the dilute gas with the macro-variables we discussed above. As we have seen, this example is an instance of the Ehrenfest-Afanassjewa condition. However, it is *not* an instance of the AET. More specifically, it is *not* the case that all sums of possible values of the one-component variable are possible values of the macro-variable f (because of the requirement that the total energy is constant, only certain sums of values of the one-component variable are possible macro-values). Hence the condition that the macro-variable K is the sum of the one-component variable where all sums of possible values of the one-component variable are possible values of the macro-variable is violated.

To conclude, the Ehrenfest-Afanassjewa condition instead and the Khinchin condition and the conditions of the AET provide sufficient but not necessary conditions. So they just identify two cases where the Boltzmannian equilibrium values and Gibbsian phase averages agree. We suspect that there will be other conditions where the Boltzmannian equilibrium values and Gibbsian phase averages agree.

4.8 Conclusion

We have considered Ehrenfest and Ehrenfest-Afanassjewa's argument for the conclusion that Boltzmannian equilibrium values and Gibbsian phase averages agree. We pointed out that their argument is true only under special circumstances. This is not a shortcoming of their proof but an inherent limitation of the claim: it is not generally the case that Boltzmannian equilibrium values and Gibbsian phase averages agree. We discussed the example of the six-vertex model and showed that in that model the two values come apart. We then offered a general theorem providing conditions for the equivalence of Boltzmannian equilibrium values and Gibbsian phase averages. The conditions of the theorem are sufficient but not necessary. This raises the important question under what other conditions Boltzmannian equilibrium values and Gibbsian phase averages agree.

References

Baxter, R. J. (1982). *Exactly solved models in statistical mechanics*. London: Academic Press.

Boltzmann, L. (1877). Über die Beziehung zwischen dem zweiten Hauptsatze der mechanischen Wärmetheorie und der Wahrscheinlichkeitsrechnung resp. den Sätzen über das Wärmegleichgewicht. *Wiener Berichte*, *76*, 373–435.

Davey, K. (2009). What is Gibbs's canonical distribution? *Philosophy of Science*, *76*, 970–983.

Ehrenfest, P., & Ehrenfest-Afanassjewa, T. (1959). *The conceptual foundations of the statistical approach in mechanics*. Ithaca, N.Y.: Cornell University Press.

Frigg, R. (2008). A field guide to recent work on the foundations of statistical mechanics. In D. Rickles (Ed.), *The ashgate companion to contemporary philosophy of physics* (pp. 99–196). London: Ashgate.

Frigg, R., & Werndl, C. (2019). Can somebody please say what Gibbsian statistical mechanics says? *The British Journal for the Philosophy of Science*, online first, https://doi.org/10.1093/bjps/axy057.

Jeans, J. H. (1904). *The dynamical theory of gases*. Cambridge: Cambridge University Press.

Khinchin, A. I. (1960) [1949]. *Mathematical foundations of statistical mechanics*. Mineola/NY: Dover Publications.

Lavis, D. (2005). Boltzmann and Gibbs: An attempted reconciliation. *Studies in History and Philosophy of Modern Physics*, *36*, 245–73.

Lavis, D. (2008). Boltzmann, Gibbs and the concept of equilibrium. *Philosophy of Science*, *75*, 682–96.

Lavis, D., & Bell, G. M. (1999). *Statistical mechanics of lattice systems, Volume 1: Closed form and exact solutions*. Berlin and Heidelberg: Springer.

Malament, D., & Zabell, S. L. (1980). Why Gibbs phase averages work. *Philosophy of Science 47*, 339–49.

Myrvold, W. C. (2016). Probabilities in statistical mechanics. In C. Hitchcock & A. Hájek (Eds.), *The Oxford handbook of probability and philosophy* (pp. 573–600). Oxford: Oxford University Press.

Slater, J. C. (1941). Theory of the transition in KH^2PO^4. *Journal of Chemical Physics*, *9*, 16–33.

Uffink, J. (2007). Compendium of the foundations of classical statistical physics. In J. Butterfield & J. Earman (Eds.), *Philosophy of physics* (pp. 923–1047). Amsterdam: North Holland.

Werndl, C., & Frigg, R. (2015a). Rethinking Boltzmannian equilibrium. *Philosophy of Science*, *82*, 1224–35.

Werndl, C., & Frigg, R. (2015b). Reconceptionalising equilibrium in Boltzmannian statistical mechanics. *Studies in History and Philosophy of Modern Physics*, *49*, 19–31.

Werndl, C., & Frigg, R. (2017a). Boltzmannian equilibrium in stochastic systems. In M. Michela & R. Jan-Willem (Eds.), *Proceedings of the EPSA15 conference* (pp. 243–254). Berlin and New York: Springer.

Werndl, C., & Frigg, R. (2017b). Mind the gap: Boltzmannian versus Gibbsian equilibrium. *Philosophy of Science*, **84**, 1289–1302.

Werndl, C., & Frigg, R. Forthcoming. When does a Boltzmannian equilibrium exist?. In D. Bedingham, O. Maroney, C. Timpson (Eds.), *Quantum foundations of statistical mechanics*. Oxford: Oxford University Press.

Werndl, C., & Frigg, R. (2020). When do Gibbsian phase averages and Boltzmannian equilibrium values agree? Studies in History and Philosophy of Modern Physics, DOI:https://www.sciencedirect.com/science/article/abs/pii/S1355219820300903.

Chapter 5
Ehrenfest and Ehrenfest-Afanassjewa on the Ergodic Hypothesis

Patricia Palacios

Abstract Ehrenfest and Ehrenfest-Afanassjewa's seminal article on statistical mechanics highlighted a crucial assumption at the heart of Boltzmann's statistical mechanics: the ergodic hypothesis. The importance of this article for transmitting the problems related with the ergodic hypothesis has been widely recognized, but Ehrenfest and Ehrenfest-Afanassjewa have been strongly criticized for not having provided a fair account of Boltzmann's statistical mechanics. In this chapter, I outline Ehrenfest and Ehrenfest-Afanassjewa's treatment of the ergodic hypothesis and I evaluate the role of this discussion for the development of the ergodic theory in the 20th century. I will conclude that the major contribution of Ehrenfest and Ehrenfest-Afanassjewa comes precisely from what has been regarded by some historians of science as historical inaccuracies of the article.

5.1 Introduction

In the *Conceptual Foundations of the Statistical Approach to Mechanics* (Ehrenfest and Ehrenfest-Afanassjewa 1959)—also known as the "Encyklopädie article"—Ehrenfest and Ehrenfest-Afanassjewa gave a prominent role to what they dubbed as "the ergodic hypothesis", suggesting that Boltzmann's entire program lacks a firm foundation because it relies on this hypothesis of questionable validity. The importance of Ehrenfest and Ehrenfest-Afanassjewa's article for transmitting the problems related with the ergodic hypothesis has been widely recognized, but they have been strongly criticized for not having provided a fair account of Boltzmann's statistical mechanics. It has been argued that they exaggerated the role of the ergodic hypothesis in Boltzmann's program (Brush 1967, 1971; von Plato 1991); that they exaggerated the importance of the equivalence between phase averages and time averages (Brush 1967; Uffink 2007); and that they misunderstood the meaning of the ergodic hypothesis as conceived by Boltzmann (von Plato 1991; Brush 1967). In this chapter, I will analyze the discussion of Ehrenfest and Ehrenfest-Afanassjewa around the ergodic

P. Palacios (✉)
Department of Philosophy (KGW), University of Salzburg, Salzburg, Austria
e-mail: patricia.palacios@sbg.ac.at

J. Uffink et al. (eds.), *The Legacy of Tatjana Afanassjewa*, Women in the History of Philosophy and Sciences 7, https://doi.org/10.1007/978-3-030-47971-8_5

hypothesis and I will evaluate the role of this discussion for the development of the ergodic theory in the twentieth century. I will conclude that the major contribution of the *Encyklopädie* article comes precisely from what has been regarded by some historians of science as historical inaccuracies of the article. In particular, I will argue that they advanced Boltzmann's interpretation of probabilities as time averages by emphasizing the role of the ergodic hypothesis and by highlighting the importance of the equivalence between phase and time averages.

This chapter is organized as follows. In Sect. 5.2, I review the origins of the ergodic hypothesis in Boltzmann's statistical mechanics, by highlighting the connection between the ergodic hypothesis and Boltzmann's time average interpretation of probabilities. In Sect. 5.3, I discuss Ehrenfest and Ehrenfest-Afanassjewa's criticism of the ergodic hypothesis and I point out that one of their major contributions was to pose a new puzzle in the foundations of statistical mechanics, which I call "the Ehrenfest and Ehrenfest-Afanassjewa's puzzle". Subsequently, in Sect. 5.4, I illustrate how this puzzle encouraged the development of the impossibility theorems in 1913. In Sect. 5.5, I argue that this puzzle also played a role in the introduction of the notion of metric transitivity, which led to the establishment of the ergodic theorems by Birkhoff and von Neumann. I point out that although these theorems solved the Ehrenfest and Ehrenfest-Afanassjewa's puzzle, they transformed the problem of ergodicity into the problem of proving that the systems of interest are metrically transitive. Finally, in Sect. 5.6, I review the recent discussion in the foundations of statistical mechanics around the problem of metric transitivity.

5.2 The Origin of the Ergodic Hypothesis

Consider a typical situation of a dilute gas enclosed in a finite container with N identical polyatomic molecules, each with r degrees of freedom. The molecules collide with each other and with the walls of the container and the collisions are governed by short-range repelling potentials. The possible states of this system are represented by points in a 2rN-dimensional phase space Γ, with q position coordinates and p momentum coordinates. Assume that the energy of the system is E, so that the state must lie on the energy surface Γ_E, which is $2rN - 1$ dimensional. At time t, the state of the system will be determined exactly by the simultaneous position and momentum coordinates of the N molecules:

$$q_1^1, \ldots, q_r^1; q_1^2, \ldots, q_r^2; \ldots, ; q_1^N, \ldots, q_r^N,$$

$$p_1^1, \ldots, p_r^1; p_1^2, \ldots, p_r^2; \ldots, ; p_1^N, \ldots, p_r^N$$

The corresponding changes in the states of the gas model are expressed, following Ehrenfest and Ehrenfest-Afanassjewa (1959)'s notation, by the following Hamiltonian equations of motion:

$$\frac{\mathrm{d}q_s^k}{\mathrm{d}t} = \frac{\delta E}{\delta p_s^k} \qquad \frac{\mathrm{d}p_s^k}{\mathrm{d}t} = \frac{\delta E}{\delta q_s^k}.$$

One can easily observe that the number of variables is enormous, so even if the system is deterministic, it is not possible to know the exact initial conditions of the system and there is little chance of integrating these equations to find the exact solutions. Such pragmatic difficulties motivated a statistical approach to the study of these kinds of systems, which began with Maxwell and Boltzmann in the second half of the ninteenth century.

Maxwell (1860) was the first to characterize the equilibrium state of a gas by a probability distribution function f. Almost a decade later, Boltzmann (1868) derived this probability distribution in the presence of external forces suggesting that if the system is left alone, the probability of the molecular velocities will *always* assume Maxwellian distribution.[1] To derive this result he uses, for the very first time, a time average interpretation of these probabilities, whereby the probability of the equilibrium state is identified as the relative time in which the system is in that state when left alone for "a very long time". In a short communication about Maxwell's work in 1879, he refers to his own interpretation of probabilities in the following terms:

> There is a difference in the conceptions of Maxwell and Boltzmann in that the latter characterizes the probability of a state by the average time in which the system is in this state, whereas the former assumes an infinity of equal systems with all possible initial states. (quoted in von Plato 1991, p. 71)

But what exactly is a time average and how does it help interpret probabilities in statistical mechanics? Let us recall that Boltzmann (1868) had adopted Maxwell's characterization of equilibrium in terms of stationary probability distributions, where macroscopic observables correspond to phase averages over the phase space with respect to a specific probability measure, i.e., the microcanonical measure. In a modernized notation, this can be written as follows:

$$\langle f \rangle = \int_{\Gamma_E} f(x)p(x)\mathrm{d}x, \tag{5.1}$$

where $f(x)$ is the phase function that gives the value of the observable for each microscopic state in the phase space, $p(x)$ is the probability distribution density, and x represents the state of the system written in local coordinates. Since this average is independent of the dynamics of the system there is no immediate interpretation

---

[1] Boltzmann noted that there might be exceptions to his derivation, for example, when the trajectory is periodic. However, he believed that such behavior would be destroyed by the slightest disturbances from outside (Uffink 2007, p. 39).

for this probability, in other words it is not clear why an arbitrarily chosen system should have this distribution. In order to establish a connection with the dynamics of the system that is robust upon the initial conditions, Boltzmann postulated that the probabilities are time averages.[2] In modern terms, this can be defined as follows. Let $T(t, x)$ represent the dynamical evolution of the system, if x is the state at time 0, then the future state at time t is $T(t, x)$. The time average of a function of state f is

$$f^* = lim_{t\to\infty}(1/t)\int_0^t f(T(t, x))\mathrm{d}t. \tag{5.2}$$

Although Boltzmann did not attempt to demonstrate the equivalence between phase averages and time averages (we will come back to this point in the next section), he did explicitly use time averages to derive the Maxwellian distribution (Brush 1967; Uffink 2007; von Plato 1991). The main strength of Boltzmann's result is its generality, indeed it is robust upon any particular assumption about collisions or any other detail of the mechanical model involved, with the only requirement that the system must obey the constancy of total energy. The main weakness of this result is that it depends on what Ehrenfest and Ehrenfest-Afanassjewa (1959) baptized as the *ergodic hypothesis*, i.e., the assumption that the trajectory of the system will eventually pass through all points of the phase space.[3] As Maxwell (1879, p. 713) puts it:

> The only assumption which is necessary for the direct proof [of the distribution law of energy] is that the system, if left to itself in its actual state of motion, will, sooner or later, pass through every phase which is consistent with the equation of energy.

Boltzmann (1868) was not unaware of the importance of this assumption for his derivation and argues: "If all initial states lead to periodic motions not running through all possible states compatible with the total energy, there would be an infinity of different possible temperature equilibria" [p. 96]. Furthermore, he was not unaware of the controversial character of this assumption. In fact, he recognized the possibility of the existence of periodic motions that fail to be ergodic. However, by studying

[2]This use of time averages for interpreting the expectation values w.r.t the Maxwellian stationary probability distribution should be distinguished from the related use of time averages to explain the empirical success of the microcanonical ensemble, which leads to the further problem of explaining that time averages are equal to the results of a macroscopic measurement. The latter is usually justified by assuming that measurements take an amount of time that is long compared to microscopic relaxation times. Since this chapter focuses on the contribution of Ehrenfest and Ehrenfest-Afanassjewa on the ergodic hypothesis, I will mostly refer to the use of time averages in the interpretation of probabilities and not to the explanation of the success of the microcanonical ensemble. For an analysis of this second problem, see e.g., van Lith (2001), Uffnk (2007), Palacios (2018).

[3]The name "ergodic hypothesis" was coined by Ehrenfest and Ehrenfest-Afanassjewa (1959). In fact, Boltzmann introduced the word *Ergode* only in 1884 to denote an ensemble of systems with a certain probability distribution in phase space. This use of the word "ergodic" by Ehrenfest and Ehrenfest-Afanassjewa has led to some controversy among commentators. For Brush (1967, p. 169), Ehrenfest and Ehrenfest-Afanassjewa gave an entire different meaning to ergodicity than the one given by Boltzmann. For Uffink (2007, p. 39) instead, they were justified in using the term "ergodic" to denote this hypothesis (see further discussion in Sect. 5.3).

the behavior of so-called Lissajous figures, he observed that small irregularities would destroy this regular behavior obliging the system to pass through every phase consistent with the equation of energy. He then justifies the ergodic hypothesis more generally by appealing to irregular distortions of external forces or surrounding gas molecules:

> The great irregularity of thermal motion and the manifold forces affecting bodies from the outside make it probable that the atoms of the warm body, through the motion we call heat, run through all the positions and velocities compatible with the equation of kinetic energy, so that we can use the equations developed above on the coordinates and component velocities of the atoms of warm bodies. Boltzmann (1871, p. 679)

It is useful to illustrate this idea with a simple example. Consider a case of a hard-sphere system in a box where every particle moves on the same straight line being reflected at each end from a perfectly smooth parallel wall. Such systems would remain on a region of the phase space without visiting the entire phase space. However, those systems would be extremely unlikely since the slightest perturbation would destroy the perfect alignment.

In spite of this, Boltzmann recognized the hypothetical character of this justification and remained skeptic about the validity of this hypothesis. This is clear by the fact that in later works (e.g., Boltzmann 1877) he attempted to characterized thermal equilibrium differently making explicit that these alternative approaches avoid the hypothesis of ergodicity (See also Uffink 2007, p. 42]).

5.3 Ehrenfest and Ehrenfest-Afanassjewa's Critique of the Ergodic Hypothesis

In their *Encyklopädie* article, Paul Ehrenfest and Tatiana Ehrenfest-Afanassjewa (1959) wrote an extensive and influential critique of Boltzmann's statistical mechanics accusing this approach of relying on what they regarded as a doubtful hypothesis, i.e., the ergodic hypothesis.

> The fundamental assumption underlying [Boltzmann's] investigation is the hypothesis that the gas models are ergodic systems [...]. With the help of this hypothesis Boltzmann computed the time average of, for instance, the kinetic energy of each atom (the same value is obtained for all atoms!). (Ehrenfest and Ehrenfest-Afanassjewa, 1959, p. 24)
>
> However, the existence of ergodic systems (i.e., the consistency of their definition) is doubtful. So far, not even one example is known of a mechanical system for which the single G-path approaches arbitrarily closely each point of the corresponding energy surface. Moreover, no example is known where the single path actually traverses all points of the corresponding energy surface (Ibid, p. 22).[4]

However, the emphasis in the doubtful character of what they called "the ergodic hypothesis" was not the main contribution of the *Encyklopädie* article. As mentioned

[4]They define a single G-path as the trajectory of the moving image point corresponding to the phase changes of a gas model (Ehrenfest and Ehrenfest-Afanassjewa 1959, footnote 73).

above, Boltzmann (1871) had already recognized the controversial character of this assumption. The main contribution of the *Encyklopädie* article, as I see it, comes precisely from what has been regarded by some historians of science as historical inaccuracies of the article (Brush 1967, 1971; von Plato 1991). In particular, that they exaggerated the role that ergodicity played in Boltzmann's statistical mechanics and that they attributed to Boltzmann an assumption that he probably never believed in. I will argue next that these aspects, which go beyond Boltzmann's investigations, are precisely the ones that really advanced the development of the ergodic theory in the twentieth century.

Although there is no consensus among modern commentators about the role and status of the ergodic hypothesis in Boltzmann's approach, most of them agree that the ergodic hypothesis was a justification for the time average interpretation of probabilities (von Plato 1991; Uffink 2007; Brush 1967). Based on what has been said in the previous section, a straightforward way of giving a time average interpretation of probabilities is by equating time averages f^* and phase averages $\langle f \rangle$:

$$\int_{\Gamma_E} f(x)p(x)\mathrm{d}x = lim_{t\to\infty}(1/t)\int_0^t f(T(t,x))\mathrm{d}t \tag{5.3}$$

However, as modern commentators have pointed out (von Plato 1991; Uffink 2007), this particular motivation for assuming ergodicity is not to be found anywhere in Boltzmann's writings and seemed to have been introduced for the first time by Ehrenfest and Ehrenfest-Afanassjewa in their review of Boltzmann's statistical mechanics. von Plato (1991, p. 78) expresses this as follows:

> This simplified reading of Boltzmann is due to the review in 1911 of foundations of statistical mechanics by Paul and Tatiana Ehrenfest. They called the above justification for assuming a single trajectory [the equivalence between phase and time averages] the "Boltzmann-Maxwell justification", and it has since then been accepted as standard. But that particular motivation for assuming ergodicity is not used by Boltzmann.

Similarly, Uffink (2007, p. 42) claims:

> There is however *no* evidence that Boltzmann ever followed this line of reasoning neither in the 1870s, nor later. He simply never gave any justification for equating time and particle averages, or phase averages, at all. Presumably, he thought nothing much depended on this issue and that it was a matter of taste.

Now since Boltzmann attributes himself a time average interpretation of probabilities, one may wonder why he did not explicitly attempt to equate phase and time averages. One can speculate that since he was aware of the possibility of periodic motions, he did not believe that this equivalence holds exactly (von Plato, 1991, p. 78). One can also believe, as Uffink (2007) does, that for him nothing much depended on this equivalence. If this latter interpretation is correct, then one should give Ehrenfest and Ehrenfest-Afanassjewa all the credit for having emphasized the important role of this equivalence for a time average interpretation of probabilities. Indeed, for them this equivalence was not just "a matter of taste" but the only way of warranting the uniqueness of the stationary probability distribution, which means that all motions associated with the same total energy yield the same value for the time average of any function (Ehrenfest and Ehrenfest-Afanassjewa, 1959, p. 22). Their reasoning can be summarized as follows. If one identifies the ergodic hypothesis as the assumption that if a system that is left to itself will pass through all the phase points compatible with its total energy. Then, given that a point in phase space cannot lie on more than one trajectory, all systems with the same value of the total energy will follow the same trajectory and their averages over infinite time intervals will be equal. The equivalence between phase and time averages gives in this way a neat interpretation of probabilities in equilibrium statistical mechanics and a clear connection to the dynamics of the system that does not depend on the initial conditions. Another significant advantage of equating time and phase averages is that phase averages can be calculated in many cases, whereas the time averages cannot (Moore 2015). The problem, as Ehrenfest and Ehrenfest-Afanassjewa stated it, is that this equivalence seems to rely crucially on the ergodic hypothesis, which they intuitively believed was not only doubtful but mathematically impossible, but more on this later.

From what has been said here, one can see that the *Encyklopädie* article, perhaps by exaggerating the role that ergodicity played in Boltzmann's approach advanced a time average interpretation of probabilities and prompted at least three important challenges that were later conceived as "the ergodic problem". The first challenge was to demonstrate that the limit involved in the definition of time averages f^* exists. The second challenge was to prove that this limit is independent of x and equal to the phase average. The third challenge was to determine the validity of the ergodic assumption or any analogous assumption required to derive this equivalence. As it will be seen later, the first two problems had to await the ergodic theorems of 1931 and 1932 and the new concept of metric transitivity (that replaced ergodicity) to be given a definite solution (See Sect. 5.5).[5] The third problem does not have an unequivocal answer yet, but the impossibility theorems of 1913 demonstrated that at least the strict ergodic hypothesis was not a valid assumption (See Sect. 5.4).

Another aspect of the *Encyklopädie* article that has been regarded as a historical error concerns the definition of the ergodic hypothesis (Brush 1967). It is important to note that Boltzmann never (at least not explicitly) associated what Ehrenfest and Ehrenfest-Afanassjewa defined as ergodic hypothesis with the word *Ergode*, which was actually introduced only in Boltzmann (1884), and was used to denote a stationary ensemble (i.e., the microcanonical ensemble) with only one integral of motion, its total energy. However, Boltzmann (1884) did assume that every element of such an ensemble traverses every phase point with the given energy. According to Uffink (2007, Footnote 20) this would have justified Ehrenfest and Ehrenfest-Afanassjewa in using this term in their formulation of the hypothesis. But beyond the discussion of whether or not the "ergodic hypothesis" should receive that name, Ehrenfest and Ehrenfest-Afanassjewa have been strongly criticized for attributing to Boltzmann

[5] Physicists frequently identify the term "ergodicity" with "metric transitivity", however, I will argue below that these are two well-defined different concepts that should not be confused.

an hypothesis that he probably never believed in. More to the point, Ehrenfest and Ehrenfest-Afanassjewa defined the ergodic hypothesis as follows:

Definition 5.1 *Ergodic hypothesis* The single, undisturbed motion of the system, if pursued without limit in time, will finally traverse "every phase point" which is compatible with its given energy. (Ehrenfest and Ehrenfest-Afanassjewa, 1959, p. 21)

For Brush (1967) and von Plato (1991), this definition is misleading for at least two reasons. First, because it refers to one single trajectory instead of an ensemble in which there is a continuous number of different trajectories. Second, because it uses the terms "every phase point", which is something that most probably Boltzmann never believed in. Regarding the first point, it is true that Boltzmann introduced the word *Ergode* to denote ensembles instead of single trajectories. However, Boltzmann did not introduce the notion of ensemble until 1884, whereas the hypothesis that the system will pass through "every phase" consistent with its given energy was used much earlier. For instance, in 1871 he explicitly refers to the motion of a point-mass in a plane under the influence of an attractive force. He says that if the force is described by a potential function $1/2(ax^2 + by^2)$ (the compound harmonic motion which results in the so-called Lissajous figures) and the ratio of the periods of the two motions is irrational, then the point-mass goes through all possible positions within a certain rectangle (Brush (1967), p. 169). Therefore, the formulation of Boltzmann's hypothesis in terms of single trajectories seems to reflect at least the first uses of this hypothesis by Boltzmann. One should also note that Ehrenfest and Ehrenfest-Afanassjewa had a further reason to formulate this hypothesis in terms of single trajectories, since, as mentioned above, by the uniqueness of mechanical trajectories, there would be essentially one trajectory, so that one can replace a long trajectory of a single system by an average over all points on the energy surface.

The second objection to Ehrenfest and Ehrenfest-Afanassjewa's formulation of the ergodic hypothesis was that it is stronger than the hypothesis that Boltzmann probably had in mind. Although Boltzmann explicitly used the terms "through every point", Brush (1967) pointed out that what he probably meant was something close to what Ehrenfest and Ehrenfest-Afanassjewa termed as the "quasi-ergodic" hypothesis, which they define as follows (footnote 98):

Definition 5.2 *Quasi-Ergodic hypothesis* The single, undisturbed motion of the system, if pursued without limit in time, will approach "arbitrarily closely each point" which is compatible with its given energy, which means that the trajectory is dense.

The reason that Brush (1967) and other historians of science (e.g., Borel 1915; von Plato 1991) offer for concluding that Boltzmann interpreted "the ergodic hypothesis" in the sense of the "quasi-ergodic hypothesis" is that in some passages, he actually added the qualification "approximation" to the description of ergodic behavior without pointing out a distinction between these behaviors and what we could understand as strict ergodic behaviors. For example, in a discussion of Kelvin's test-cases in 1892 he claims: "all possible sets of values of x, y, and θ which are consistent with the equation of *vis viva* are obtained with any required degree of approximation [*mit beliebiger Annherung erreicht werden*]" (quoted in Brush (1967), p. 173). They suppose then that the use of the terms "through every point" were just used by Boltzmann as an approximation of trajectories going "arbitrarily close to every point" (Brush (1967), p. 174). We could express this idea in terms of contemporary philosophy of science and say that Boltzmann believed strict ergodic trajectories to be a *straightforward idealization* of the trajectories going arbitrarily close to every point, i.e., an idealization that constitutes an approximation of realistic behavior and that can therefore be de-idealized without loosing explanatory power (See Butterfield (2011)).

If the previous interpretation is correct, then it seems then that Ehrenfest and Ehrenfest-Afanassjewa were wrong in taking the idea of trajectories going literally "through every point" in Boltzmann's writings too seriously, after all Boltzmann just meant trajectories going "arbitrarily close to every point" (Borel 1915; von Plato 1991; Brush 1967). However, this aspect, which appears to be a historical misconception of the *Encyklopädie* article and the result of "some careless statements made by the Ehrenfests" (Brush 1967, p. 169), is at the same time one of the main contributions of the article about the ergodic hypothesis. As I see it, Ehrenfest and Ehrenfest-Afanassjewa advanced Boltzmann's ideas by pointing out that the difference between "ergodic behavior" and "quasi-ergodic behavior" is mathematically essential, which implies that the ergodic hypothesis cannot be de-idealized and replaced by the quasi-ergodic hypothesis without loosing explanatory power. In footnote (98) they illustrate this essential difference between "ergodic" and "quasi-ergodic" behavior by help of the following example. Consider a geodesic line of a torus for which the ratio of the two numbers of turnings in the two directions is irrational. Such a geodesic intersects the meridian at infinitely many points P_h, which are densely distributed everywhere over the circumference. No matter how many times one turns around the torus along the geodesic line, one will never get from a point P_h to the diametrically opposite point Q on the meridian. This is because if that were the case, then twice the same number of revolutions would bring us back to P_h and the system would stay in a particular region of the phase space, thus failing to describe a dense trajectory. This means that the points that can be visited by a dense trajectory constitute a subset of all points of the phase space. "From this one can easily see that the set of all those points P_h which can be reached by a given geodesic line form a denumerable subset in the continuum of all those points on the circumference which the geodesic line approaches arbitrarily closely" (Ehrenfest and Ehrenfest-Afanassjewa, 1959, Footnote 98). Here Ehrefest and Ehrenfest-Afanassjewa were not only pointing out an essential difference between "ergodic" and "quasi-ergodic" trajectories, but also suggesting that the notion of strict ergodic behavior was not consistent and therefore mathematically impossible (See Sect. 5.4).

But there was a further and more important reason to distinguish between "ergodic" and "quasi-ergodic" behavior, namely, that they suspected that quasi-ergodic behavior does not entail the desired conclusion that Maxwell's distribution is the only stationary distribution over the energy surface and therefore that it does not warrant the equivalence between phase and time averages: "[W]e must say that for a "quasi-ergodic" system on each surface $E(q, p)$ there will be a continuum of $\infty^{(2rN-2)}$ different G-paths with different values of the constants $c_2, \ldots, c_{2rN_1}$. Hence one cannot extend the Boltzmann-Maxwell justification [...] to quasi-ergodic systems" (footnote 99). "The time average in question can change quite discontinuously from path to path for a quasi-ergodic system, because we obtain it by averaging over an infinite time interval" (footnote 102).

Although they did not offer a careful proof of the previous statements, their distinction between ergodic and quasi-ergodic hypothesis suggests that the ergodic hypothesis corresponds to what in the contemporary philosophy of science (Fletcher et al. 2019) is called an "essential idealization", i.e., an idealization that cannot be de-idealized without loosing explanatory power. In this case, the idealized assumption was the ergodic hypothesis, which Ehrenfest and Ehrenfest-Afanassjewa suggested cannot be de-idealized by the weaker and more plausible "quasi-ergodic hypothesis". The latter ideas led them to the following puzzle:

> *Ehrenfest and Ehrenfest-Afanassjewa's puzzle*: On the one hand, by assuming ergodicity, one can demonstrate the equivalence between phase and time averages. Yet the existence of ergodic systems is doubtful. On the other hand, if one de-idealizes this assumption by the weaker hypothesis of quasi-ergodicity, which is probably true for some systems, one does not obtain the desired equivalence.

This clear way of presenting the issues surrounding the ergodicity assumption is in my view the second major contribution of the *Encyklopädie* article on this topic, apart from the above-mentioned emphasis on the equivalence between time and phase averages. We will see in the next sections that in order to solve the Ehrenfest and Ehrenfest-Afanassjewa's puzzle, one needs to introduce elements of modern topology and the new notion of metric transitivity, which, as I will argue below, corresponds neither to the ergodic hypothesis nor to the quasi-ergodic hypothesis.

5.4 Proof of the Impossibility of Ergodic Systems

We have seen above that the *Encyklopädie* article not only raised suspicion on the actual existence of ergodic systems but also on the mathematical possibility of ergodic systems. The challenge was therefore not only for physicists, who were asked to justify the validity of this assumption but also for mathematicians, who were now challenged to demonstrate that ergodic systems cannot exist in principle. Mathematicians Arthur Rosenthal and Michel Plancherel accepted the challenge and independently demonstrated in 1913, soon after the publication of the *Encyklopädie* article, that Ehrenfest and Ehrenfest-Afanassjewa's intuition was correct and that a mechanical system represented by a phase space with more than one dimension cannot pass through every point on the energy surface. Rosenthal (1913) explicitly recognizes the direct influence of Ehrenfest and Ehrenfest-Afanassjewa in his "Proof of the impossibility of ergodic gas" and begins his article by saying: "In view of the fact that no example of such an ergodic system has been demonstrated with certainty

P. and T. Ehrenfest doubted the existence of ergodic systems (i.e., they doubted that their definition is not contradictory). In the following it will be shown that this doubt was correct; i.e., *it will be shown* that not only *no gas* is an ergodic system, but also that in general such systems cannot exist." (p. 796)

In order to understand the proofs offered by Rosenthal (1913) and Plancherel (1913), it is necessary to review some of the results obtained in pure mathematics during the second half of the nineteenth century, in which these proofs were based. The ergodic hypothesis, as formulated by Ehrenfest and Ehrenfest-Afanassjewa, implies that a certain curve, which can be placed in correspondence with a straight line (the time axis), eventually visits all points of the phase space. This means that such a curve would appear one dimensional from the time axis point of view and multidimensional from the phase space point of view, if the phase space has more than one dimension. This obliged to introduce a method for comparing the sizes of infinite classes of different dimensionalities. Cantor, who developed his work on set theory during 1871 and 1897, furnished such a method by using the criterion of one-to-one correspondence. The consequence of Cantor's theory that was relevant for the understanding of ergodicity was the proof that any n-dimensional manifold can be put into one-to-one correspondence with any m-dimensional manifold, where n and m may have different dimensionalities (Cantor 1878).[6] However, this proof lacks the property of continuity, which means that points that are close together in the n-dimensional manifold may be mapped into points that are far apart in the m-dimensional manifold. Cantor suspected then that bicontinuous one-to-one mapping may serve as a criterion for proving that two sets have the same dimensionality, which was finally demonstrated by Brouwer (1911).

Other developments needed to establish the impossibility of ergodic systems in 1913 were the results offer by Borel (1898) and Lebesgue (1902), who completed Cantor's method for comparing infinite sets of points by providing a method for determining the length, area, volume, and more generally, the "measure" of sets of points. In these approaches, the measure of a point is defined to be zero whereas the measure of all real numbers in a finite interval is defined as the length of that interval. Based on these results, Plancherel (1913) and Rosenthal (1913) published their celebrated proofs. Without going into technical details, Rosenthal (1913)'s proof can be summarized as follows. Consider a gas system of N particles and r degrees of freedom, where all states of the system are represented by a $2rN$-dimensional phase space Γ. One can choose a small region G of the energy surface Γ_E in which all the partial derivatives of the energy with respect to the coordinates and momenta are continuous functions of the coordinates and momenta, such that at least one of these derivatives is different from zero everywhere in that region. One can then map this region G into a $2rN-1$-dimensional cube. According to the ergodic hypothesis, the representative point of the system must pass through every point in that region G. This can happen in two ways: (i) In a finite time interval (one-dimensional time axis), the representative point enters G, visits all points, and comes out again, which means that the $2rN-1$-dimensional region G should be mapped onto a line of

[6] See Brush (1967) for a historical review of these results.

finite length, continuously and one-to-one. (ii) Or the representative point passes into and out of G infinitely many times. Rosenthal's proof shows that neither of these alternatives is possible, except in the trivial case when the phase space is one dimensional. For both alternatives imply that there is a bicontinuous one-to-one mapping in sets of different dimensionalities, which according to Brouwer's proof is impossible. Plancherel's proof follows a similar reasoning but using Lebesgue theory of measure. In particular, the proof is based on the result that the time axis, a line, is a set of measure zero with respect to a region of two or more dimensions.[7]

These proofs demonstrated that the first part of what has been called here "Ehrenfest and Ehrenfest-Afanassjewa's puzzle" was correct: a mechanical system cannot be ergodic, except in cases when the phase space is one dimensional. It remained to be demonstrated the truth of the second part of this puzzle, namely, that the weaker "quasi-ergodic" hypothesis was not sufficient to derive the equivalence between time and phase averages. An indirect proof of this statement had to await the establishment of the ergodic theorems of Birkhoff (1931) and von Neumann (1932), which stated the necessary and sufficient conditions for the equivalence between time and phase averages. We review these results in the following section.

5.5 The Ergodic Theorems and the Notion of Metric Transitivity

In the 1930s, Birkhoff (1931) and John von Neumann (1932) published two separated papers containing different versions of what is now known as "the ergodic theorem". This theorem provided a key insight into the problem presented in a clear way for the first time by Ehrenfest and Ehrenfest-Afanassjewa, namely, the justification of the hypothesis that time averages equal phase averages. At the same time, it initiated an entire new field of mathematical research called ergodic theory, which has thrived more than 80 years.[8]

The basic concept that allowed von Neumann and Birkhoff to arrive at the celebrated result was the notion of "metric transitivity", introduced for the first time by Birkhoff and Smith (1928). In order to understand these results and the notion of metric transitivity, one needs to become familiarized with elements of measure theory. Let (X, ϕ_t, μ) be a dynamical system given by a metric space (phase space) X, a continuous map $\phi : X \to X$, and equipped with a normalized Lebesgue-measure μ (i.e., $\mu = 1$) restricted to X. A point $x \in X$, which represents a particular state of the system, "moves" in time, generating a "flow" that we denote as $\phi_t(x)$. $\phi_t(x)$ is the position to which the system moves after time t so that $\phi_t(x)$ is the solution of the differential equation with initial value x at time $t = 0$. One should also add that $\phi_t(x)$ is a homeomorphism of X onto itself, which satisfies the group property $\phi_t\phi_s = \phi_{t+s}$.

[7] See Brush (1967) for more details about this proof.

[8] See Mackey (1974) for an excellent historical review of the ergodic theory.

It also preserves the Liouville measure μ. The notion of "metric transitivity", which replaced the old notion of "ergodicity" introduced by Boltzmann (1868), was then defined as follows:

Definition 5.3 *Metric transitivity* The dynamical system (X, ϕ_t, μ) is *metrically transitive* iff for any measurable set of nonzero measure V and for *almost every point* $x \in X$ it holds that $\{\phi_t(x)\} \cap V \neq \emptyset$ for some time t.

This means that eventually almost every point $x \in X$ (i.e., all points except for a set of measure 0) visits every measurable set V in X. If a dynamical system is metrically transitive, it follows that it cannot be decomposed into two (or more) invariant regions of nonzero measure.

By using this notion of metric transitivity, the theorems of Birkhoff (1931) and von Neumann (1932) established that the time limit $t \to \infty$ used in the following definition of time averages f^* exists for almost all x and is independent of x when it exists:

$$f^* = lim_{t\to\infty}(1/t) \int_0^t f(\phi_t(x))dt, \tag{5.4}$$

where f is an integrable function on the phase space X, representing a physical measurement on a system that is in state $x \in X$.[9]

In short, they demonstrated that the following theorem (the Ergodic Theorem) holds:

Theorem 5.1 Ergodic Theorem *If the dynamical system (X, ϕ_t, μ) is metrically transitive, then the limit of f^* exists and coincides with the phase average $\langle f \rangle = \int_X f(x)d\mu(x)$, for* almost all $x \in X$.

The proof of Theorem 5.1 constituted the first crucial step toward the solution of the long-standing problem of the equivalence between phase and time averages posed in a clear way by Ehrenfest and Ehrenfest-Afanassjewa. However, as one can observe, this theorem depends essentially on the assumption that systems are "metrically transitive", an assumption that is not quite easy to justify. In this sense, one can say that the ergodic theorem transformed the question of equivalence between time and phase averages into the question of whether the flow ϕ representing the time evolution of the system is metrically transitive. I will return to this problem in the next section, but first I will put these results in the context of the previous discussion around the ergodic hypothesis.

In Sect. 5.3, we said that Ehrenfest and Ehrenfest-Afanassjewa posed the following puzzle: On the one hand, if we assume ergodicity, then one can derive the equivalence between time averages and phase averages. However, real systems cannot be ergodic. On the other hand, if we assume the weaker hypothesis of quasi-ergodicity,

[9]von Neumann (1932) demonstrated that the functions of x on the time average converge and Birkhoff (1931) proved further that this convergence was pointwise almost everywhere. See Moore (2015) for more details on the difference between von Neumann and Birkhoff's results.

which may be true of some systems, then one cannot derive the equivalence between phase and time averages. In Sect. 5.4, we saw that the first part of this puzzle was correct, indeed Rosenthal (1913) and Plancherel (1913) demonstrated that real systems represented by a phase space with more than one dimension cannot be ergodic. This naturally raises the question of whether metric transitivity is equivalent to the original ergodic hypothesis. If they were equivalent, then Theorem 5.1 would loose interest, since it would not be applicable to almost any real system of interest. Fortunately, the notion of metric transitivity has been proven to be weaker than the original ergodic hypothesis and therefore immune to the impossibility proofs of ergodic systems elaborated by Rosenthal and Plancherel (2015). The key aspect that weakens the definition of metric transitivity is the introduction of the expression "for almost every x" in the definition of metric transitivity, which means for all x except for a set of measure zero.[10] Furstenberg (1961) demonstrated by considering an r-dimensional torus that convergence "almost everywhere" can be replaced by "convergence everywhere" only in cases of $r = 1$, which is consistent with Rosenthal and Plancherel's results. In all other cases, i.e, for $r > 1$, one needs to impose further restrictions on the transformation ϕ.

Another question that arises in light of the Ehrenfest and Ehrenfest-Afanassjewa's puzzle is whether the notion of metric transitivity corresponds to what they dubbed as the "quasi-ergodic" hypothesis. If this were the case, then the second part of the puzzle would have been proven to be false, since "quasi-ergodicity" would have been demonstrated to be sufficient for the equivalence between phase and time averages. Interestingly, although metric transitivity is frequently taken as equivalent to the quasi-ergodic hypothesis (e.g., Lebowitz and Penrose (1937)), the former can be demonstrated to be stronger than the later.[11] More specifically, as Ehrenfest and Ehrenfest-Afanassjewa define it, the quasi-ergodic hypothesis corresponds to the hypothesis that the orbits (trajectories) are topologically dense in the phase space (i.e., they pass arbitrarily close to every point of phase space). It can be demonstrated that metric transitivity implies quasi-ergodicity, when X is a compact metric, which is generally assumed in the ergodic theorems (Moore 2015; Kolyada and Snoha 1997). However, the converse is not valid, since it requires the further assumption that X has no isolated point, which does not necessarily hold in the systems of interest (Kolyada and Snoha 1997).[12] In fact, it is not even true that a minimal flow with an invariant measure, in which every orbit is dense, is metrically transitive (Moore 2015). Since metric transitivity is a necessary and sufficient condition for the validity

[10]There is an interesting foundational problem associated with the definition of metrical transitivity for "almost every x", which is called "the measure zero problem". The issue is that it is hard to demonstrate that states with probability measure zero can be neglected without begging the question, i.e., without presupposing that phase averages equal time averages. I will not discuss this problem further for lack of space, but the reader can see Uffink (2007), van Lith (2001), Frigg (2016) for a detailed discussion around this issue.

[11]Perhaps the misleading title of von Neumann's paper (1932) "Proof of the quasi-ergodic hypothesis" contributed to this confusion.

[12]Systems in which X has no isolated point is said to be a standard dynamical system (Kolyada and Snoha 1997).

of Theorem 5.1., then one can conclude that the "quasi-ergodic hypothesis" is too weak to establish the equivalence between phase and time average, which means that the second part of the Ehrenfest and Ehrenfest-Afanassjewa's puzzle was also true.

Let us now summarize the contribution of Ehrenfest and Ehrenfest-Afanassjewa to the development of the ergodic theorems in the 1930s. The first direct contribution was to point out for the very first time the importance of the hypothesis that time averages equal phase averages in a time average interpretation of probabilities. In absence of this emphasis on the role of this hypothesis, there may have not been enough motivation to provide such careful mathematical proofs of this equivalence. Another direct contribution of Ehrenfest and Ehrenfest-Afanassjewa was to suggest that neither the ergodic hypothesis nor the quasi-ergodic hypothesis can serve to derive the desired equivalence between time and phase averages for the systems of interest, which we have called here the Ehrenfest and Ehrenfest-Afanassjewa's puzzle. This motivated mathematicians and physicists to find a different hypothesis, i.e., metric transitivity, that can serve as a basis to derive the equivalence between phase and time averages. The proof of this "ergodic" theorem based on the notion of metric transitivity finally solved the puzzle prompted two decades before by Ehrenfest and Ehrenfest-Afanassjewa. But, as said above, it led to a different, yet more specific problem: the problem of demonstrating that the real physical systems of interest are in fact metrically transitive. In the next section, I will discuss this issue further.

5.6 The Problem of Metric Transitivity

How can one prove that the flow of a dynamical system is metrically transitive? And how can we be sure that metrically transitive systems exist? These have proved to be very challenging questions that have motivated an enormous amount of research (e.g., Oxtoby and Ulam 1941; Markus and Meyer 1947; Sinai 1970). A promising result of the existence of metrically transitive systems was offered Oxtoby and Ulam (1941), who showed that on a compact polyhedron equipped with a finite Lebesgue measure, all measure-preserving homeomorphisms are metrically transitive in a topological sense. A more concrete example of metrically transitive systems was examined by Sinai (1970), who considered a model of dynamical systems, where the molecules were contained in a cubical enclosure and moved with periodic boundary conditions. Although the model was not entirely realistic (it allowed for collisions between the molecules but not with the walls of the container), he proved that the system was (approximately) metrically transitive. These results contrasted with the ones offered by Markus and Meyer (1947), who showed that for Hamiltonian dynamical systems, almost all systems fail to be metrically transitive. The latter was reinforced by the so-called KAM (Kolmogorov, Arnold, Moser) Theorem, which stated that when the interactions among the molecules are non-singular, the phase space will contain islands of stability where the flow is not metrically transitive (see Lichtenberg and Lieberman 2013; Earman and Rédei 1996). To be more specific, the theorem shows that if one starts by a Hamiltonian system with quasi-periodic trajectories and adds perturbation terms that are intended to eliminate this periodic behavior, there will still remain "islands" of periodic behavior so that the system fails to be metrically transitive. Based on these results one can conclude that most systems of interest in statistical mechanics are very probably not metrically transitive (Wightman 1985; Earman and Rédei 1996; van Lith 2001).

Different reactions can be found among philosophers of science on the consequences of these results. For some (e.g., Earman and Rédei 1996; van Lith 2001) these results lead to the conclusion that the traditional ergodic (or metrically trasitivity) program for interpreting probabilities and explaining the success of phase averaging should be abandoned. Others have suggested instead (Vranas 1998; Frigg and Werndl 2011) that the ergodic program can be rescued by appealing to what they call "epsilon-ergodicity" (or more precisely "epsilon-metrical transitivity"). This latter view is motivated by the possibility that most systems that fail to be metrically transitive have a probability measure that is *close enough* to the microcanonical. More specifically, (Vranas 1998) examines computational evidence for the existence of systems that he calls "epsilon-ergodic", which are systems that have an invariant subset B of measure $1 - \epsilon$, such that i) ϵ is tiny or zero and ii) for *almost every point* $x \in X$ it holds that $\{\phi_t(x)\} \cap B \neq \emptyset$ for some time t. Then, by generalizing the notion of absolute continuity, he proved that if a system is "epsilon-ergodic" (which here can be understood as "epsilon-metrically transitive"), the probability measure associated to that system will be close to the microcanonical. One can build an intereresting parallel with the previous discussion around the ergodic hypothesis and interpret this "epsilon-ergodicity hypothesis" as an analog of the quasi-ergodic hypothesis. The hope is then that one can de-idealize the hypothesis of metrical transitivity by the hypothesis of "epsilon-ergodicity" or "epsilon-metrical transitivity" without loosing explanatory power, since it is expected that in the latter case the probability measure will be *close enough* to the microcanonical measure.

Even accepting Vranas' results, there still remains the question of whether most of systems of interest are "epsilon-ergodic" or more precisely "epsilon-metrically transitive". There is important evidence suggesting the existence of systems that display thermodynamic behavior and yet are not metrically transitive or even "epsilon-metrically transitive" (Frigg 2016 A field guide to recent work on the foundations of statistical mechanics Uffink 2007). For example, in a solid the molecules can oscillate around fixed positions so that the phase trajectory of the system can only access a small part of the energy hypersurface (Uffink 2007; Frigg 2016). Frigg and Werndl (2011) have argued that this is not as problematic as it seems, since one can still use the ergodic theory for the restricted set of cases that are proven to be "epsilon-ergodic" (epsilon-metrically transitive) such as gases. However, Earman and Redei (1996) are skeptic about this line of reasoning, since they claim that if there are non-metrically transitive systems that display thermodynamic-like behavior, then it is likely that the same mechanisms that explain this behavior in non-metrically transitive systems also explain the behavior of metrically transitive systems (or epsilon-ergodic) systems.[13]

[13]Werndl and Frigg (2015) prove a theorem that establishes that for equilibrium to exist three factors need to cooperate: the choice of macro-variables, the dynamics of the system, and the choice of the

The question about the validity of the ergodic theory based on the notion of metric transitivity that followed the discussion started by Ehrenfest and Ehrenfest-Afanassjewa remains open in the foundations of statistical mechanics. One of the reasons why this theory has not been given up despite the complications to demonstrate that systems are in fact metrically transitive and other problems associated with this approach is that this theory gives a solid foundation to the time average interpretation of probabilities and a neat mechanical explanation of thermodynamic equilibrium (Frigg 2016).[14] Giving up the time average interpretation of probabilities in order to get rid of the problems associated with the ergodic hypothesis is a high price to pay, since the alternatives face similar if not more dramatic problems. Indeed, the frequentist interpretation of probabilities violates the requirement of von Mises's theory (van Lith 2001; Frigg 2016), and the propensity interpretation (Popper 1959) is inconsistent with the assumption of a deterministic underlying micro theory (Clark and Butterfield 1987; Frigg 2016). A different strategy to deal with the problem of interpreting probabilities consists in avoiding probabilities all in all. This approach is known as the "typicality approach" and is based on the distinction between "typical states", which correspond to equilibrium states and "atypical" states, which are non-equilibrium states (Lebowitz 1993; Goldstein 2001). Although this program solves some problems associated with the ergodic theory, it has been criticized for not establishing a clear connection with the dynamics of the system (Frigg 2009, 2010). Very recently, (Wallace 2016) has suggested a novel interpretation of statistical mechanical probabilities based on quantum-mechanical probabilities, yet the empirical validity of this approach is still to be seen.

5.7 Conclusion

We have traced the history of the ergodic hypothesis from its origins to recent discussions in the foundations of statistical mechanics highlighting the contribution of Ehrenfest and Ehrenfest-Afanassjewa in this debate. We have seen that apart from pointing out the difficulties associated to the demonstration of the existence of ergodic systems, they motivated a distinction between "ergodic" and "quasi-ergodic" systems and, more importantly, they emphasized the role of the hypothesis that time

state space. For them a consequence of this theorem is that focusing on ergodicity as the crucial property for the existence of an equilibrium state is misleading.

[14]There are other important problems associated with the ergodic approach that has not been mentioned here because they are not directly related with the problems discussed by Ehrenfest and Ehrenfest-Afanassjewa. One of these problems is the measure zero problem, which I mentioned on footnote 9. The other problem concerns the justification of infinite time limits in the definition of time averages. The worry here is that if one wants to explain the empirical success of the microcanonical distribution it is not clear why one should interpret measurements as time averages, even less as infinite time averages (Uffink 2007; Frigg 2016; Palacios 2018). Finally, there is the problem that time average interpretation cannot easily be generalized to time-dependent phenomena and therefore this theory seems to be restricted to the explanation of equilibrium and cannot be used as a general non-equilibrium theory (See Uffink 2007; van Lith 2001).

averages equal phase averages in a time average interpretation of probabilities. We have seen that the latter, which has been sometimes regarded as a historical error of their analysis, served to the postulation of the ergodic theorem in the 1930s and encouraged the development of the concept of metric transitivity, which continues playing a role in the foundations of statistical mechanics.

References

Birkhoff, G. D. (1931). Proof of the ergodic theorem. *Proceedings of the National Academy of Sciences*, *17*(12), 656–660.

Birkhoff, G. D., & Smith, P. (1928). Structure analysis of surface transformations. *Journal de Mathématiques pures et appliquées*, *7*, 345–380.

Boltzmann, L. (1868). Studien über das Gleichgewicht der lebenden Kraft zwischen bewegten materiellen Punkten. *Wiener Berichte*, *58*, 517–560.

Boltzmann, L. (1871). Einige allgemeine Sätze über Wärmegleichgewicht. *Wiener Berichte*, *63*, 679–711.

Boltzmann, L. (1877). Über die Beziehung zwisschen dem zweiten Haubtsatze der mechanischen Wärmetheorie und der Wahrscheinlichkeitsrechnung resp. dem Sätzen über das Wärmegleichgewicht. *Wiener Berichte*, *76*, 373–435.

Boltzmann, L. (1884). Über die Eigenschaften monozyklischer und anderer damit verwandter Systeme. *Crelle's Journal für die reine und angewandte Mathematik*, *98*, 68–94.

Borel, E. (1898). *Leçons sur la théorie des fonctions*. Paris: Gauthier-Villars.

Borel, E. (1915). *Mécanique Statistique. Exposé d'aprés l'article allemand de P. Ehrenfest, T. Ehrenfest*. Paris: Gauthier-Villars.

Brouwer, L. E. (1911). Beweis der invarianz der Dimensionenzahl. *Mathematische Annalen*, *70*(2), 161–165.

Brush, S. G. (1967). Foundations of statistical mechanics 1845–1915. *Archive for History of Exact Sciences*, *4*(3), 145–183.

Brush, S. G. (1971). Proof of the impossibility of ergodic systems: The 1913 papers of Rosenthal and Plancherel. *Transport Theory and Statistical Physics*, *1*(4), 287–298.

Butterfield, J. (2011). Less is different: Emergence and reduction reconciled. *Foundations ofphysics*, *41*(6), 1065–1135.

Cantor, G. (1878). Ein beitrag zur Mannigfaltigkeitslehre. *Journal für die reine und angewandte Mathematik*, *84*, 242–258.

Clark, P., & Butterfield, J. (1987). Determinism and probability in physics. *Proceedings of the Aristotelian Society, Supplementary Volumes*, *61*, 185–243.

Earman, J., & Rédei, M. (1996). Why ergodic theory does not explain the success of equilibrium statistical mechanics. *The British Journal for the Philosophy of Science*, *47*(1), 63–78.

Ehrenfest, P., & Ehrenfest-Afanassjewa, T. (1959). *The conceptual foundations of the statistical approach to mechanics* Cornell University Press, Ithaka. English translation. In F. Klein (Ed.), *M.J. Moravcsik ofBegriffliche Grundlagen der statistischen Auffassung in der Mechanik*. Encyklopädie der mathematischen Wissenschaften IV-32 (pp. 1–90, 1911). Re-used by Dover, Minneola (2015).

Fletcher, S. C., Palacios, P., Ruetsche, L., & Shech, E., (2019). Infinite idealizations in science: An introduction. *Synthese*, *196*(5), 1657–1669.

Frigg, R. (2009). Typicality and the approach to equilibrium in boltzmannian statistical mechanics. *Philosophy of Science*, *76*(5), 997–1008.

Frigg, R. (2010). Why typicality does not explain the approach to equilibrium. In *Probabilities, causes and propensities in physics*. Springer, Berlin.

Frigg, R. (2016). A field guide to recent work on the foundations of statistical mechanics. In *The Ashgate companion to contemporary philosophy of physics*. Routledge.

Frigg, R., & Werndl, C. (2011). Explaining thermodynamic-like behavior in terms of epsilon-ergodicity. *Philosophy of Science*, *78*(4), 628–652.
Furstenberg, H. (1961). Strict ergodicity and transformation of the torus. *American Journal of Mathematics*, *83*(4), 573–601.
Goldstein, S. (2001). Boltzmanns approach to statistical mechanics. In *Chance in physics*. Springer, Berlin.
Kolyada, S., & Snoha, L. (1997). Some aspects of topological transitivity a survey. *Grazer Mathematische Berichte*, *334*, 3–35.
Lebesgue, H. (1902). Intégrale, longueur, aire. *Annali di Matematica Pura ed Applicata (1898–1922)*, *7*(1), 231–359.
Lebowitz, J. L. (1993). Macroscopic laws, microscopic dynamics, time's arrow and Boltzmann's entropy. *Physica A: Statistical Mechanics and its Applications*, *194*(1–4), 1–27.
Lebowitz, J. L., & Penrose, O. (1973). Modern ergodic theory. *Physics Today*, *26*(2), 23–29.
Lichtenberg, A. J., & Lieberman, M. A. (2013). *Regular and stochastic motion*. Springer Science & Business Media.
Mackey, G. W. (1974). Ergodic theory and its significance for statistical mechanics and probability theory. *Advances in Mathematics*, *12*(2), 178–268.
Markus, L., & Meyer, K. R. (1974). *Generic Hamiltonian dynamical systems are neither integrable nor ergodic*. Memoirs of the American Mathematical Society 144.
Maxwell, J. C. (1860). Illustrations of the dynamical theory of gases. *The London, Edinburgh, and Dublin Philosophical Magazine and Journal of Science*, *19*(124), 19–32.
Maxwell, J. C. (1879). On Boltzmann's theorem on the average distribution of energy in a system of material points. *Transactions of the Cambridge Philosophical Society*, *12*, 547–570.
Moore, C. C. (2015). Ergodic theorem, ergodic theory, and statistical mechanics. *Proceedings of the National Academy of Sciences*, *112*(7), 1907–1911.
Neumann, J. V. (1932). Proof of the quasi-ergodic hypothesis. *Proceedings of the National Academy of Sciences*, *18*(1), 70–82.
Oxtoby, J. C., & Ulam, S. M. (1941). Measure-preserving homeomorphisms and metrical transitivity. *Annals of Mathematics*, 874–920.
Palacios, P. (2018). Had we but world enough, and time... but we dont!: Justifying the thermodynamic and infinite-time limits in statistical mechanics. *Foundations of Physics*, *48*(5), 526–541.
Plancherel, M. (1913). Beweis der Unmöglichkeit ergodischer mechanischer Systeme. *Annalen der Physik*, *347*(15), 1061–1063.
Popper, K. R. (1959). The propensity interpretation of probability. *The British Journal for the Philosophy of Science*, *10*(37), 25–42.
Rosenthal, A. (1913). Beweis der Unmöglichkeit ergodischer Gassysteme. *Annalen der Physik*, *347*(14), 796–806.
Sinai, Y. G. (1970). Dynamical systems with elastic reflections. Ergodic properties of dispersing billiards. *Uspehi Mat Nauk*, *25*(2), 141–192.
Uffink, J. (2007). Compendium to the foundations of classical statistical physics in handbook for the philosophy of physics. In *Handbook for the philosophy of physics*. Amsterdam: Elsevier.
van Lith, J. (2001). Ergodic theory, interpretations of probability and the foundations of statistical mechanics. *Studies in History and Philosophy of Science Part B: Studies in History and Philosophy of Modern Physics*, *32*(4), 581–594.
von Plato, J. (1991). Boltzmann's ergodic hypothesis. *Archive for History of Exact Sciences*, *42*(1), 71–89.
Vranas, P. B. (1998). Epsilon-ergodicity and the success of equilibrium statistical mechanics. *Philosophy of Science*, *65*(4), 688–708.
Wallace, D. (2016). Probability and irreversibility in modern statistical mechanics: Classical and quantum. In *Quantum foundations of statistical mechanics*. forthcoming: Oxford University Press.

Werndl, C., & Frigg, R. (2015). Reconceptualising equilibrium in Boltzmannian statistical mechanics and characterising its existence. *Studies in History and Philosophy of Science Part B: Studies in History and Philosophy of Modern Physics*, *49*, 19–31.
Wightman, A. S. (1985). Regular and chaotic motions in dynamical systems: Introduction to the problems. In *Regular and chaotic motions in dynamic systems*. New York: Plenum.

Chapter 6
The Ehrenfests' Use of Toy Models to Explore Irreversibility in Statistical Mechanics

Joshua Luczak and Lena Zuchowski

Abstract This article highlights and discusses the Ehrenfests' use of toy models to explore irreversibility in statistical mechanics. In particular, we explore their urn and P–Q models and highlight that, while the former was primarily used to provide a simple counter-example to Zermelo's objection to Boltzmann's statistical mechanical underpinning of the Second Law of Thermodynamics, the latter was intended to highlight the role and importance of the *Stoßzahlansatz* as a cause of the tendency of systems to exhibit entropy increase. We also explain the sense in which these models are toy models and why agents can use them, as the Ehrenfests' did, to carry out this important work, despite the fact that they do not represent any real system.

6.1 Introduction

Several contributions to this volume have demonstrated that Tatiana Afanassjewa was not just a brilliant mathematical physicist but also a methodological and didactical innovator. This paper will focus on a methodological aspect of the work on statistical mechanics she conducted jointly with her husband, Paul Ehrenfest, which again shows their willingness to approach known material from surprising angles: namely, their use of toy models to explore irreversibility.

The Ehrenfests' urn and P–Q model have become legendary and are often used as didactical tools in lectures on statistical mechanics. However, until very recently, the literature on scientific modelling paid scant attention to such toy models. Building on the investigation into the functions of toy models by one of the authors (Luczak 2017), we are now able to investigate the Ehrenfests' toy models in this framework and show that, not only do they provide an excellent illustration of how toy models work in general but they also illustrate a particular advantage of this viewpoint. Namely, we will argue that their non-representational status allows toy models to be

J. Luczak
School of Social Science, Singapore Management University, Singapore, Singapore

L. Zuchowski (✉)
Department of Philosophy, University of Bristol, Bristol, England
e-mail: lena.zuchowski@bristol.ac.uk

J. Uffink et al. (eds.), *The Legacy of Tatjana Afanassjewa*, Women in the History of Philosophy and Sciences 7, https://doi.org/10.1007/978-3-030-47971-8_6

used complementary to investigate aspects of a given system and that the interplay between the urn and P–Q model (and the later Kac-model) provides a good example of such use. To our knowledge, both the complementary use of toy models in general, as well as the specific use of the Ehrenfests' ones, have not been investigated in detail yet.

In Sect. 6.2, we will discuss toy models, their typical functions, and their unique properties. In Sect. 6.3, the urn model (Sect. 6.3.1) and the P–Q model (Sect. 6.3.2) will be described. Section 6.4 will describe why and how these models work as a means of investigating aspects of a given target system. In Sect. 6.5, we will then discuss the complementary use of several toy models. Conclusions will be drawn in Sect. 6.6.

6.2 Toy Models and Their Functions

A scientific model is an object, which may be either real or abstract, that has a certain set of properties, and is used by agents for various scientific, modelling activities and purposes. Many of these activities are aimed at developing our knowledge or understanding of the physical world. For example, scientists frequently use models:

1. To test the compatibility of various concepts (i.e. in a consistency proof).
2. To elucidate certain ideas relevant to a theory. That is, to reach a clearer understanding of an idea, its implications, and its relation to other ideas within a theory.

Scientists often use simple models in ways such as these because it is typically easier to work with an object that instantiates fewer properties than the kinds of physical systems we want to better understand.

What is interesting to note, however, is that in a subset of situations in which scientists use models in ways such as these, they write or speak about their models in such a way so as to convey, with more or less emphasis, that they do not intend them to, at least on these occasions, perform a representational function. This renders their models nonrepresentational. Despite this, their models nonetheless help to develop a better understanding of the world. They are nonetheless helpful because they instantiate properties that are shared by the kinds of systems we want to understand.

Let us begin by laying down an expression that is intended to capture the kinds of models, uses and intentions that are important for our purposes. Let us use the expression "toy models", and say that toy models are simple models that are not intended to perform a representational function but rather to perform some other important function, such as 1 and 2. We recognise that the label, "toy model", is perhaps a little bit awkward since the expression is sometimes used to refer to simple models that might otherwise be called highly or strongly idealised (see, for example, Alexander & Dominik Forthcoming and Jamesh 2019) but, since we wish to discuss some situations in which some simple models are being used nonrepresentationally, we will use the expression "toy models" for those cases, and want to suggest saving expressions such as "highly idealized model" or "strongly idealized model" for simple models that are at least performing a representational function. Regardless, that is, of how we want to understand what makes them representational or idealised.

As we highlight in the coming sections, Tatiana Afanassjewa and Paul Ehrenfest (according to the convention used in this book, we will refer to them as 'the Ehrenfests' in the following) developed a collection of simple models that performed a number of important functions—including those listed above. Since there is little reason to believe that these models are intended to represent any actual system and because agents can successfully use them in precisely the ways the Ehrenfests did without intending that they perform a representational function, it is reasonable to regard their models as toy models. We will treat their models as toy models in the reminder of this chapter.

While it is a subject of debate within philosophy of science as to what exactly constitutes a model's representation of a target, all of the leading substantive accounts of scientific representation agree that a necessary condition of a model's representation of a target is that its user intends that it perform a representational function.[1] The inclusion of this condition not only fits with the promising and growing view that scientific representation is a practice performed by intentional agents but it also helps to ensure that what a model represents, if it represents, is not ambiguous and that analyses of scientific representation account for the logical properties of scientific representation.[2] Scientific representation is neither a reflexive or symmetric relation. Models do not (typically) represent themselves. When they perform a representational function, they represent something else. Be it a particular system or some collection of systems. And when they represent other systems, those other systems are not (at least typically thought to be) representations of the model. By including the intentions of model users, analyses of scientific representation obtain the required asymmetry and irreflexivity. Users typically do not intend that targets represent their models nor that their models represent themselves. Since toy models are models that are not intended to perform a representational function, it is the case that they do not perform a representational function.

Of course, since toy models do not perform a representational function, one may wonder about their relevance is for science. More pointedly, one may wonder how they can perform important functions, such as 1 and 2, if they do not perform a representational function. Models, such as the urn and P–Q model, can be used by agents to do a lot of interesting and important work simply because either they instantiate certain properties or because they instantiate certain properties that are also known to be instantiated by other systems.

In the latter type of case, these similarities permit treating the model as an analogue. Moreover, these similarities permit, and are sufficient for, analogical reason-

[1] See, for example, Frigg and Nguyen (2016) and Frigg and Nguyen (2017).

[2] Historically, however, intentionality conditions have usually been added to analyses of scientific representation only after they have been rejected for failing to account of the logical properties of scientific representations.

ing.[3] That is, they permit, and are sufficient for, employing some version of the following argument schema, where S is some model and T is some other system:

P1. S is similar to T in certain (known) respects.
P2. S has some further feature Q.
C. Therefore, T also has the feature Q, or some feature Q^* similar to Q.

Importantly, since similarity is a reflexive and symmetric relation, it cannot be the case that it is sufficient for scientific representation. So then, even if a model instantiates properties that are also instantiated by other systems, this does not entail that it represents any or all of those systems or anything at all.

In the following section, we intend to highlight the use of the urn and P–Q model in a consistency proof (use **1**), and their use in elucidating the relationship between an important statistical mechanical concept, irreversibility, and an important statistical mechanical assumption, the *Stoßzahlansatz* (use **2**).[4] In Sect. 6.4 we discuss why agents can use these models, as the Ehrenfests did, to successfully perform these tasks, despite the fact that they do not represent any real system. We also discuss why it is appropriate to regard the urn model as primarily aimed at providing a simple counter-example to Zermelo's objection to Boltzmann's statistical mechanical underpinning of the Second Law of Thermodynamics and why it is appropriate to regard the P–Q model as primarily aimed at highlighting the role and importance of the *Stoßzahlansatz* as a cause of the tendency of systems to exhibit entropy increase.

6.3 The Ehrenfests' Toy-Models

In this section, we will discuss two seminal toy models developed by Tatiana Afanassjewa and Paul Ehrenfest: the urn model.[5] (Sect. 6.3.1) and the P–Q model (Sect. 6.3.2). Both models explore the relationship between irreversibility and the *Stoßzahlansatz*. However, the aspects of this problem they focus on and their respective formalisms are very different.

6.3.1 The Urn Model

Purpose: The explicit purpose of the urn model is to provide a simple counter-example to Zermelo's objection to Boltzmann's statistical mechanical underpinning of the Second Law of Thermodynamics: i.e. the assertion that it seems "unimaginable"[6] that the entropy of an ensemble of particles with reversible microdynamics

[3] See Bartha (2013) for a comprehensive discussion of analogies and analogical reasoning.
[4] For a discussion of other important functions performed by toy models see Luczak (2017).
[5] The urn model was introduced in Ehrenfest and Ehrenfest (1907).
[6] unvorstellbar.

should "as a rule"[7] increase with time. To falsify this claim, the urn model provides a scenario in which such a tendency of entropy increase over time follows from elementary probability calculations about the distribution of different states in an ensemble of particles.[8] The model's microdynamics, i.e. the moving back and forth of balls between two urns, therefore is fundamentally reversible: any ball can (and if the model is run over sufficiently long time scales, will) move in both directions between the urns. However, the specific balls moving into a different urn at a given time are selected randomly. By separating the reversible and irreversible parts of the model, the Ehrenfests are able to demonstrate that reversible microdynamics in combination with a well-contained probabilistic component, i.e. an analogy to the *Stosszahlansatz*, can lead to an entropy-increase.

Model set-up: The urn model[9] consists of two urns, labelled A and B, and a bag, labelled L. Initially, urn A contains $n_0^{(A)}$ balls and urn B contains $n_0^{(B)}$ balls, where $N = n_0^{(A)} + n_0^{(B)}$ is the total number of balls. The balls carry labels n ranging from 1 ... N. The bag L, the "lottery" container, contains N consecutively numbered balls, or the same number of other objects (e.g. paper slips, wood chips), which allow for randomisation of the numbers.

At each time step t, the lottery sack L is shaken and a number n is drawn at random. The ball carrying this number "jumps"[10] from the urn it is currently in into the other one. The model can, of course, be easily realised materially.

Model behaviour: It is easy to see that, as the number of time steps t increases, the numbers of balls in each urn will tend to even out. Since equal numbers of balls in each urn corresponds to the state of maximum entropy, this implies that the entropy will show a tendency to increase as well. In the following, we will briefly outline the Ehrenfests' probabilistic argument for this increase.[11]

Suppose that, at time t, urn A contains $n_t^{(A)}$ balls and urn B contains $n_t^{(B)}$ balls. The probabilities that a ball is chosen to jump from urn A to urn B and the other way around are given by, respectively:

$$p_t^{(A\rightarrow B)} = \frac{n_t^{(A)}}{N} \tag{6.1}$$

$$p_t^{(B\rightarrow A)} = \frac{n_t^{(B)}}{N} \tag{6.2}$$

The difference between those probabilities is then directly proportionally to the difference in the number of balls in each urn:

[7] in der Regel.

[8] See Ehrenfest and Ehrenfest (1907, p. 2).

[9] For the original description of the set-up, see Ehrenfest and Ehrenfest (1907, pp. 2–4).

[10] hüpft.

[11] See Ehrenfest and Ehrenfest (1907, p. 2–3).

$$\Delta p = \frac{n_t^{(A)} - n_t^{(B)}}{N}, \tag{6.3}$$

i.e. it will be more likely ($\Delta p > 0$) that a ball moves from urn A to urn B if there are currently more balls in urn A than in urn B ($n_t^{(A)} > n_t^{(B)}$), and more likely ($\Delta p < 0$) that a ball moves from urn B to urn A if there are currently more balls in urn B than in urn A ($n_t^{(A)} < n_t^{(B)}$). Accordingly, the model tends to balance out the numbers of balls in each urn. This does not imply, of course, that lottery draws that increase the difference in those numbers may never occur. However, the larger the existing difference in the numbers of balls is, the less likely such draws become:

> It is always more probable that the chosen ball is in the fuller rather than in the emptier urn. Therefore, while urn A is still much fuller than urn B, urn A will empty itself into [urn] B during a sequence of draws and only seldom receive a ball from [urn] B.[12]

Moral: The Ehrenfests conclude that the toy model demonstrates that, if a micro-dynamics is assumed that treats all microstates as equally accessible by all particles, then the tendency of an ensemble of particles towards maximum entropy follows from elementary probability considerations.[13] However, they concede that the assumption of an equiprobable state-space distribution ("molecular chaos"),[14], i.e. the *Stoßzahlansatz*, which is represented in the urn model through the random draws from the lottery sack, is a premise of the model. The toy model should, therefore, be seen as demonstrating the consequences of the *Stoßzahlansatz* rather than as providing justification for this assumption itself:

> We are therefore not touching on the question, in how far the proof of the [Second Law of Thermodynamics] can be seen as complete; which specific meaning one should give the hypothesis of "permanent molecular chaos".[15]

Further development by Kac: A modified version of the Ehrenfests' (1907) urn model was presented by Kac.[16] Since this toy model was already discussed in detail by one of the authors,[17] we will only briefly outline the major differences between the Kac's ring model and the urn model. Rather than representing particle dynamics as the random movements of balls between two urns, the ring model represents these dynamics as rotations and colour changes on a ring of balls.

[12]Ehrenfest and Ehrenfest (1907, p. 2): Es ist immer wahrscheinlicher, daß die jeweils aufgerufene Kugel in der volleren, als daß sie in der leereren Urne angetroffen wird. Solange also die Urne A noch viel voller ist als die Urne B, wird sich bei den folgenden Ziehungen in der Regel die Urne A in B entleeren und nur ausnahmsweise eine Kugel aus B erhalten.

[13]See Ehrenfest and Ehrenfest (1907, p. 4).

[14]molekulare Unordnung.

[15]Ehrenfest and Ehrenfest (1907, p. 4, footnote 1): Deshalb lassen wir hier die Frage durchaus unberührt, inwieweit der Nachweis des H-Theorems etwa als lückenlos angesehen werden kann; welchen Sinn man im speziellen der Hypothese einer dauernden "molekularen Unordnung" geben soll.

[16]See e.g. Kac (1956).

[17]See Luczak (2017).

In particular, the model consists of N black and white balls placed onto a circle and connected to each other by a corresponding number of N-edges. A number of $N_2 \leq N$ markers are then randomly distributed over the edges and the balls are moved step-wise along the edges of the circle. Furthermore, whenever a ball traverses an edge with a marker it changes colour. The set-up of the ring-model, therefore, possesses one crucial difference to the urn model: the randomising *Stoßzahlansatz* only needs to be invoked once, i.e. during the placing of the markers. Therefore, any dynamics on the ring are fully reversible and the colour development of the balls shows additional recurrent properties (if the model is observed for a sufficient number of time steps). Nevertheless, every single ring will—for the majority of its development—tend towards equal numbers of black and white balls, i.e. to a state of maximum entropy. Furthermore, for an ensemble of rings with randomly chosen N_2, the average entropy development will monotonously increase for a period of $2N$ time steps.

However, given that Kac's ring model still includes an explicit implementation of the *Stoßzahlansatz*, the conclusions to be drawn from this model appear to be essentially identical to the ones drawn from the original urn model: both toy models demonstrate that—assuming a randomising component like the *Stoßzahlansatz* is included in the dynamics—a tendency to increase entropy follows from elementary probabilistic calculations on the distribution of particles through the appropriate state space.

6.3.2 *The P–Q model*

In this section, we will discuss the purpose, formalism and behaviour of the P–Q model.[18] The model is also known as the 'wind-tree model'. However, this seems to be something of a misnomer since the model is clearly intended as a model of the interaction of two ensembles of particles.

Purpose: The discussion of the P–Q model is labelled as an "interlude".[19] The Ehrenfests describe the purpose of the model as the provision of a demonstration of "the place the *Stoßzahlansatz* occupies in the Maxwell-Boltzmann investigation [of the tendency of entropy increase]".[20] In contrast to the urn model, the P–Q model is therefore not primarily a means to demonstrate the likelihood of a tendency towards entropy increase but serves to highlight the importance of the Stoßzahlansatz as a cause of this tendency. It therefore advances the conceptual investigation that was started with the urn model (Sect. 6.3.1): the earlier toy model aims to show that it is possible to have a model combining reversible microdynamics with a probabilistic component that shows a statistical increase in entropy; the later investigates the the

[18] The P–Q model was introduced in Ehrenfest and Ehrenfest (1909).

[19] Zwischenstück.

[20] Ehrenfest and Ehrenfest (1909, p. 19): ... welche Stellung der *Stoßzahlansatz* in den zuletzt erwähnten *Maxwell-Boltzmannschen* Untersuchungen einnimmt.

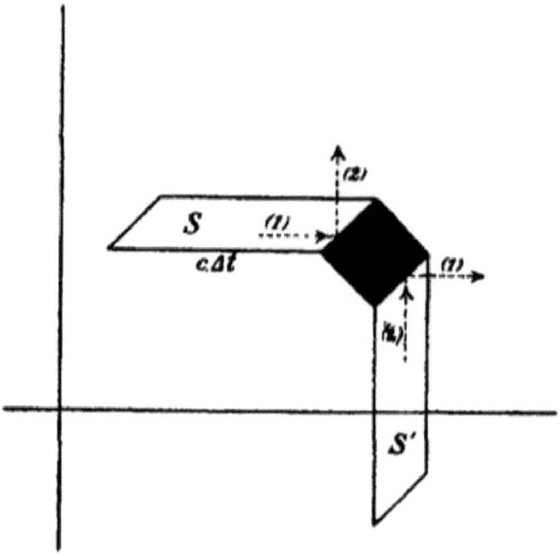

Fig. 6.1 Interaction between P- and Q-molecules in the P–Q model. Original illustration from Ehrenfest and Ehrenfest (1909, p. 20)

precise place of the probabilistic component in a scenario with collision dynamics (rather than the abstract urn movements of the urn model). In particular, the probabilistic component is now restricted to the assumption that the 'collision sites' are distributed through the available space in a way that makes it equally likely for each molecule of a given population to experience a collision. The P–Q model, therefore, has a much more direct implementation of the *Stoßzahlansatz*, showing that randomising the number of collisions ('Stoesse') in sufficient to inducing an overall increase of entropy.

Model set-up: The P–Q model[21] is introduced as a highly-simplified model of the interactions between two populations of molecules: the P-Molecules and the Q-Molecules. The P-Molecules are point-particles, which move frictionless and without experiencing external forces on a plane. The P-Molecules do not collide with each other and can only move in the four principal directions (along the directions of the $-x$-, x-, $-y$- and x-axis) of a given coordinate system inscribed on the plane.

In contrast to the P-Molecules, the Q-Molecules are motionless squares of length a. The diagonals of the squares are aligned with the x- and y-axis, i.e. the four sides of the Q-Molecules are oriented at angles of $\phi = \pi/4$ to all possible directions of movement of the P-Molecules (see Fig. 6.1). The Q-Molecules are distributed uniformly over the plane: each "larger area contains approximately the same [number of Q-Molecules]".[22] Furthermore, the average distance between two Q-Molecules is much greater than their principal length a.

---

[21] For a description of the model's set-up, see Ehrenfest and Ehrenfest (1909, pp. 19–20).

[22] Ehrenfest and Ehrenfest (1909, p. 19): ... auf jedes grösseres Gebiet sollen nahe gleichviel entfallen ...

Let N be the number of P-Molecules and $n_t^{(-x)}$, $n_t^{(x)}$, $n_t^{(-y)}$ and $n_t^{(y)}$ the number of P-Molecules moving in the direction of the $-x$-, x-, $-y$- and y-axis at time t, respectively. Due to the relative orientation of Q-Molecules' sides and the P-Molecules' directions of movement, an elastic collision between a P- and a Q-Molecule will change the former's direction of movement by deflecting it into one of the orthogonal directions of movement: e.g. after a collision with a Q-Molecule, a P-Molecule originally moving in the $-x$-direction will either move in the $-y$- or the y-direction. The sign of the direction of movement after a collision depends on which of the sides of the Q-Molecule the P-Molecule hits (see Fig. 6.1).

At time t, the number of molecules changing from a given direction i to an orthogonal direction j during the next time step Δt is then given by:

$$n_{t,\Delta t}^{(i\to j)} = k_{i,j,\Delta_t} n_t^{(i)}, \tag{6.4}$$

where k_{i,j,Δ_t} is the fraction of P-molecules moving in direction i that experience a collision with a side of a Q-molecule such that they are deflected towards direction j.

The *Stoßzahlansatz* is implemented into the model through the following simplifying assumption about the fractions of molecules experiencing collisions k_{i,j,Δ_t}:

> The analogy to the *Stoßzahlansatz*, which we mentioned several times before, now consists in the following *claim*:
>
> The fraction of P-Molecules with each direction of movement in the band S is such that it corresponds to the ratio of the total area of all bands S to the total free space available.[23]

Thereby, the bands S are finite projections of the relevant side of a given Q-Molecule, i.e. they constitute a kind of catchment area of the Q-Molecule's sides, so that, within time step Δt, all P-Molecules within this areas which are moving in the projected direction i will experience a collision that reflects them to direction j (see Fig. 6.1). Since the implementation of the *Stoßzahlansatz* prescribes that the fraction of molecules experiencing such collisions is proportional to the total area of the bands S, it is the same for all P-molecules, regardless of their direction of movement. Furthermore, since the P-Molecules are moving with constant speeds, the lengths of the bands S (and hence the ratio of their totality to the total available area) is directly proportional to the length of the time step Δt:

$$k_{i,j,\Delta t} = k\Delta t, \tag{6.5}$$

where k is a constant that depends on the side length a

Behaviour: The number of molecules changing from direction i to direction j only depends on the time step Δt and the number of molecules moving in direction i at the beginning of this time step:

---

[23] Ehrenfest and Ehrenfest (1909, p. 20): Das Analogon zu dem mehrfach genannten *Stoßzahlansatz* besteht nun in der folgenden *Behauptung*: Von den P-Molekülen jeder einzelnen Bewegungsrichtung entfällt auf die Streifen S ein solcher Bruchteil, als dem Verhältnis der Gesamtfläche aller S zur totalen freien Fläche entspricht.

$$n_{t,\Delta t}^{(i\to j)} = k\Delta t n_t^{(i)}. \tag{6.6}$$

However, the same must be true for the reverse kind of collisions, i.e. reflections from direction j to direction i:

$$n_{t,\Delta t}^{(j\to i)} = k\Delta t n_t^{(j)}. \tag{6.7}$$

The total difference between the number of molecules changing from direction i to direction j and from direction j to direction i is then given by:

$$\Delta n^{i\leftrightarrow j} = k\Delta t |(n_t^{(i)} - n_t^{(j)})| \tag{6.8}$$

Accordingly, at each time step, the effect of the collisions between Q- and P-Molecules will be such as to counteract any differences between the numbers of P-molecules moving in directions i and j. Since this computation can be generalised to any pair of direction i and j, The Ehrenfests conclude that the implementation of the *Stoßzahlansatz* leads to a monotonous approach of a uniform distribution of the numbers $n_t^{(i)}$:

> If the *Stoßzahlansatz* [(as quoted above)] is consistently applied to the calculation of the numbers [$n_{t,\Delta t}^{(j\to i)}$] during each time step Δt, then one obtains a monotonous decrease of the differences between the numbers [$n_t^{(x)}, n_t^{(-x)}, n_t^{(y)}, n_t^{(y)}$].[24]

This also implies an increase of entropy in the model and an equilibrium distribution of velocities corresponding to:

$$n_t^{(x)} = n_t^{(-x)} = n_t^{(y)} = n_t^{(y)} = \frac{N}{4}, \tag{6.9}$$

which, in this simplified model, is the equivalent of Maxwell's velocity distribution.[25]

Moral: By using a scenario that is somewhat closer to a real many-particle situation than the one represented by the urn model, the P–Q model allows the Ehrenfests to pinpoint he role of the *Stoßzahlansatz* more precisely. It highlights that the apriori assumption of an equal accessibility of all states by all particles is crucial to obtaining the dependency of the change in particle numbers on the existing velocity distribution, which underlies the tendency of the model to minimise such differences.

The P–Q model also has the advantage that—as in the Kac model—the argument for an entropy increase only involves the consideration of fully deterministic terms: the randomisation through the *Stoßzahlansatz* is accomplished through the choice for the constants $k_{i,j}$ rather than through repeated random draws.

---

[24]Ehrenfest and Ehrenfest (1909, p. 20): Wenn bei der Berechnung der Zahlen N_{12}, N_{21}, N_{23}, N_{32} etc. für jedes Zeitelement Δt immer wieder der *Stoßzahlansatz* (7) zugrunde gelegt wird, so erhält man eine monotone Abnahme der Unterschiede der Zahlen f_1, f_2, f_3, f_4.

[25]See Ehrenfest and Ehrenfest (1909, p. 19).

However, the Ehrenfests emphasise that the toy model only demonstrates a tendency of entropy increase if the *Stoßzahlansatz* is assumed as a premise: the justification of this premise needs to be provided by an unrelated argument.[26]

6.4 Why These Toy Models Work

In this section, we explain why agents can use the urn model to highlight the compatibility of various concepts, as the Ehrenfests did, despite the fact that it does not represent any real system. We also explain why agents can use the P–Q model to elucidate the relationship between irreversibility and the *Stoßzahlansatz*, again, as the Ehrenfests did, despite the fact that it too does not represent any real system. We also explain why it is appropriate to regard the urn model as primarily aimed at providing a simple counter-example to Zermelo's objection to Boltzmann's statistical mechanical underpinning of the Second Law of Thermodynamics and why it is appropriate to regard the P–Q model as primarily aimed at highlighting the role and importance of the *Stoßzahlansatz* as a cause of the tendency of systems to exhibit entropy increase. We begin with some historical background.

Ludwig Boltzmann famously attempted to account, in a classical framework, for the fact that isolated systems away from equilibrium spontaneously approach equilibrium and that they thereafter remain in equilibrium, if they are not interfered with. In 1872, Boltzmann considered how the distribution of velocities of the molecules of a contained dilute gas could be expected to change under collisions and argued that there was a unique distribution—now called the Maxwell–Boltzmann distribution—that was stable under collisions.[27] Boltzmann further argued that a gas that initially had a different distribution would move towards the Maxwell–Boltzmann distribution. To argue for this, Boltzmann defined a quantity, which we now call H, showed that it reached a minimum value for the Maxwell-Boltzmann distribution, and argued that it would *monotonically* decrease to its minimum.[28] This result is now known as Boltzmann's H-theorem. It is a straightforward consequence of Boltzmann's transport equation. Importantly, it is a temporally asymmetric result.[29]

In the wake of this result, many began to wonder how Boltzmann arrived at it, having only assumed a dynamics that is symmetric under time reversal. It was later discovered that he did not, and two famous objections have shown that he could not. These are known as the reversibility and recurrence objections. The former is usually credited to Josef Loschmidt and the latter to Ernst Zermelo.[30] The reversibility objec-

---

[26]For the discussion by the Ehrefests, see Ehrenfest and Ehrenfest (1909, pp. 18–19); for reasons to treat the *Stoßzahlansatz* as true, see Luczak (2016).

[27]See Boltzmann (1872).

[28]The quantity we call H was originally denoted E in Boltzmann's early work. See Boltzmann (1872).

[29]See Brown et al. (2009) for more on Boltzmann's H-theorem.

[30]See Uffink (2007) and Brown et al. (2009) for more on these objections.

tion applies to systems whose microdynamics are symmetric under time reversal. In the case of Boltzmann's gas, it says that for any set of trajectories of the molecules of the gas, the time-reversed trajectories are also compatible with the dynamics. So not *all* microstates of the gas at *any* time lead to a *monotonic* decrease of H. The recurrence objection applies to classical systems with bounded phase space energy hypersurfaces. That is, to systems, with total fixed energy, such as Boltzmann's gas. If we consider a small open neighbourhood of the system's initial state and ask, will the system, after it leaves that neighbourhood, ever return to it? Then the answer, which makes use of Henri Poincarè's recurrence theorem, is *yes*, it will, for almost all initial phase space-points, i.e. for all except a set of Lebesgue measure zero. More plainly, but less precisely, the objection notes that no initial microstate will yield a *monotonic* decrease of H.

To derive Boltzmann's original, asymmetric, result, one needs more than what is given by simply applying Newton's laws of motion to molecular collisions. For Boltzmann, it was a temporally asymmetric assumption that appeared in the derivation of his transport equation. The assumption, which posits an absence of correlations between the velocities of colliding molecules at all times, is now known as the *Stoßzahlansatz*.[31]

In the light of this discussion, and, in particular, the recurrence and reversibility objections, one may wonder whether it is possible to reconcile irreversible macroscopic behaviour with an underlying dynamics that is recurrent and symmetric under time reversal. It is precisely this thought that underlies the objection raised by Zermelo that we noted earlier, i.e. that it seems "unimaginable" that the entropy of an ensemble of particles with reversible microdynamics should "as a rule" increase with time. In response to this challenge, the Ehrenfests offer the urn model (Sect. 6.3.1).

Importantly, the challenge Zermelo sets ask whether certain properties are compatible—namely, are there systems which exhibit observable irreversible behaviour, whose evolution is recurrent and symmetric under time-reversal? Since these properties are perfectly general, and not tied to any particular system, the challenge can be successfully answered by locating or constructing a system, real or otherwise, that consistently instantiates these properties. Let us call the set of systems that instantiate all of these properties J. As we saw in the previous section, the urn model consistently instantiates these properties. It is a member of J. So it can be used by an agent, as it was by the Ehrenfests, to address this challenge. Of course, there is nothing unique to the urn model in this respect. Other members of J that consistently instantiate these properties would work equally well. If, in constructing and establishing this consistency proof, a user does not intend that the urn model perform a representational function, then it does not perform a representational function. It neither represents some other system, the set J of systems, or anything else. In such a situation, the urn model is a toy model. An agent can use the model in this consistency proof and answer Zermelo's challenge without intending that it per-

[31]The term "*Stoßzahlansatz*" was coined by Paul and Tatiana Ehrenfest (1907). See Uffink (2007) and Brown et al. (2009) for more on the *Stoßzahlansatz*.

form a representational function because it instantiates the set of properties whose compatibility is questioned.

While the urn model successfully addresses Zermelo's challenge, it does so by essentially incorporating the *Stoßzahlansatz* into the dynamics. This obscures the role and importance of the assumption, as one might think that this is merely an artefact of the model. What is needed then, if one wants to elucidate the relationship between irreversibility and the *Stoßzahlansatz* is to locate or construct a model, real or otherwise, whose tendency of entropy increase crucially depends on adding the assumption to the model. The most vivid way of doing this is to explicitly appeal to the assumption after specifying a dynamics that does not either directly incorporate it or have it as a consequence. This is precisely what the Ehrenfests did when they proposed their P–Q model. In effect, this move amounts to adding an additional property to J, where this property is a constraint on a system's allowable dynamics, and to abandoning the requirement that the new set, $J + S$, be consistent. Since this additional property/constraint is, like the other properties that characterise J, perfectly general, and not tied to any particular system, one can elucidate the relationship between irreversibility and the *Stoßzahlansatz* by locating or constructing a model, real or otherwise, whose tendency of entropy increase crucially depends on adding the assumption to the model after specifying a dynamics that does not either directly incorporate it or have it as a consequence. If, in elucidating the relationship between irreversibility and the *Stoßzahlansatz*, a user does not intend that the P–Q model perform a representational function, then it does not perform a representational function. It neither represents some other system, the set $J + S$ of systems, or anything else. In such a situation, the P–Q model is a toy model. An agent can use the model to elucidate the relationship between irreversibility and the *Stoßzahlansatz* without intending that it perform a representational function because it instantiates the requisite set of properties.

With these ideas having been expressed, we are now in a position to appreciate why it is appropriate to regard the urn model as primarily aimed at providing a simple counter-example to Zermelo's objection to Boltzmann's statistical mechanical underpinning of the Second Law of Thermodynamics and why it is appropriate to regard the P–Q model as primarily aimed at highlighting the role and importance of the *Stoßzahlansatz* as a cause of the tendency of systems to exhibit entropy increase. As they have been presented here, following the Ehrenfests' discussion, only the urn model consistently instantiates the properties that are elements of J. Since the *Stoßzahlansatz* is a temporally asymmetric assumption, it is, strictly speaking, incompatible with the P–Q model's symmetric dynamics.[32] So the model, so described, cannot be used to address Zermelo's challenge. It can, however, be used to address this challenge by suitably modifying it so that the *Stoßzahlansatz* is approx-

[32]This is easiest to see if we consider the extreme disequilibrium state. Suppose that, at $t = 0$, $k_{-x,j,\Delta_t} = 1$, and that the *Stoßzahlansatz* holds. A time Δ_t later, a fraction of these P-molecules have been scattered, half into the $-y$-direction, half into the y-direction. Suppose, now, we reverse the velocities, and ask what fraction of, say, the y-direction P-molecules will collide with a Q-molecule in time Δ_t. Answer: all of them! The P-molecules that are not travelling in the x-direction are all on collision courses that will turn them into x-travelling molecules.

imately true (see Appendix A of Brown et al. 2009). Since the *Stoßzahlansatz* is not an explicit assumption in the urn model, but rather, in essence, incorporated into its stochastic dynamics, no conflict arises between it and the urn model's symmetric dynamics. It can be used to address Zermelo's challenge. Since the importance and effective role of the *Stoßzahlansatz* is obscured in the urn model but vivid in the P–Q model, it is for these reasons that we say that it is appropriate to regard the P–Q model as primarily aimed at highlighting the role and importance of the *Stoßzahlansatz* as a cause of the tendency of systems to exhibit entropy increase and appropriate to regard the urn model as primarily aimed at providing a simple counter-example to Zermelo's objection.

6.5 Complementary Use of Toy Models

As explained in Sect. 6.4, the Ehrenfests' toy models do not represent any one system but foster the drawing of conclusions via arguments from analogy about the specific aspects in which they are similar to other systems and to classes of statistical mechanical systems. The fact that these models do not represent allows for the complementary use of several different toy models in an inquiry into different aspects of a system. In this respect, toy models are crucially different from models that represent: different models that represent the same target system need either be viewed as exclusive alternatives to each other or as ranked according to representative accuracy.[33] In contrast, multiple toy models, which resemble a given system in different ways, can be used complementary to each other.

This process can be formalised using the reasoning scheme introduced above (Sect. 6.2) by adding additional reasoning steps based on the presence of a second model to it. Suppose S_1 and S_2 are both toy models and T is some other system. Then we can—unproblematically and without having to worry about their precise relationship to each other—use the two models in a process of iterative reasoning:

P1. S_1 is similar to T in certain (known) respects.
P2. S_1 has some further feature Q_1.
C1. Therefore, T also has the feature Q_1, or some feature Q_1^* similar to Q_1.
P3. S_2 is similar to T in certain (known) respects, which do not need overlap, or even be compatible, with the respects in which S_1 is similar to T.
P4. S_2 has some further feature Q_2.
C2. Therefore, T also has the feature Q_2, or some feature Q_2^* similar to Q_2.
C. Therefore, T also has the features Q_1 and/or Q_2, or some feature Q_1^* and/or Q_2^* similar to Q_1 and Q_2, respectively.

The iterative investigation of T through a series of toy models $S_1, S_2, \ldots, S_n$ can be extended to any number n of steps.

[33]For a description of modelling with representing models, see e.g. Frigg and Nguyen (2016, 2017).

So far, we have not said anything about the relationship between the features of S_1 and S_2. In an unproblematic use of the iterative reasoning formalised above, these will be distinct features, so that the use of different toy models will simply allow us to explore more and more features of T. However, it might also be the case that some of the attributed features, Q_i and Q_j, are not compatible with each other, either because they describe the same feature in different ways or because the system T cannot have both features simultaneously. In this case, the inquiry needs to be extended by a second step, in which the scientist judges which feature should actually be attributed to T. The justification for this decision will likely refer to the degree of similarity the models have to aspects of the original system, as established in steps P1 and P3 above. It is notable that this verification requires considerably less than the comprehensive evaluation of rival representational models: it only requires an evaluation of the specific aspects in which the toy models are similar to T, concluding with a decision about which model is more realistic with respect to these aspects, or which model's realistically presented aspects are more crucial to the system T.

The use of the Ehrenfests' toy models (Sect. 6.3) can be seen as an example of such complementary reasoning and partial evaluation. Their investigation aims at establishing two related features Q_1 and Q_2: whether irreversible behaviour is compatible with an underlying physics that is recurrent and symmetric under time-reversal (Q_1, a feature of the urn model S_1); and how irreversible behaviour manifests from an underlying deterministic dynamics with a *Stoßzahlansatz*-like assumption (Q_2, a feature of the P–Q model S_2).

With respect to the first concern, we can point to the urn model and use it to settle the compatibility question. With respect to the second, it should be noted that we can use the urn model, in an argument from analogy, to draw the conclusion that irreversible behaviour trades on a stochastic dynamics of a particular sort. Since we have reason to believe or at least take, the underlying dynamics of ordinary statistical mechanical systems (e.g. a gas) to be deterministic, we reject this conclusion. If we move then to a model that incorporates a deterministic dynamics, say, the P–Q model, on the basis of this consideration and find that in order to exhibit irreversible macroscopic behaviour we need a *Stoßzahlansatz*-like assumption, either because it is the most simple or natural or, perhaps unbeknownst to us, the only way of achieving it, then we draw the conclusion, using an argument from analogy, that such behaviour is manifest in the gas because of the assumption. We have now moved from the initial conclusion C_1, that irreversible behaviour is compatible with a reversible dynamics, to the more precise conclusion that C_2, that it can be generated by deterministic dynamics with a *Stoßzahlansatz*-like assumption.

It is notable that the two models are not consistent with each other: the urn model has stochastic dynamics (Q_3) and the P–Q model does not. However, since we are not interested in either model being a precise representation of T, or even any representation of T, but only use these models as complementary, investigative tools, this is not problematic for us.

However, since the two features Q_3 (stochastic dynamics) and Q_2 (deterministic dynamics and *Stoßzahlansatz*) are not compatible with each other, a decision needs to be made about which feature is to be the one we conclude about T. The outline of the reasoning above pre-empts this decision: the latter feature is seen as the correct one. It is only at this point that a comparison of the resemblance of the two models is conducted: the P–Q system has all things considered, and on balance, more properties in common with everyday statistical mechanical systems, such as a gas system, and, in a number of cases, these properties are more fine-grained. For example, while both the P–Q and urn model involve changes in particle numbers, only the P–Q model involves changes that result from collisions.

The epistemological advantages gained from such the complementary use of toy models are obvious: it allows for the selective investigation and clarification of particular features, or the ability to investigate different questions related to a single topic, without having to appeal to representational models. In fact, the non-representativeness of toy models makes this complementary use possible since it allows for the models to be treated not as incompatible representations of the same scenario but as investigative tools in a common inquiry.

The Ehrenfests' use of toy models illustrates these advantages: it begins by asking two questions about the nature of irreversible behaviour. One of them is concerned with whether such behaviour is compatible with an underlying physics that is recurrent and symmetric under time reversal. The other question is concerned with how such behaviour manifests itself in a gas system from its underlying physics. Since the use of toy models relieves any worries about using incompatible models when investigating and attempting to answer different questions related to irreversible behaviour, the Ehrenfests' are free to devise and use whatever toy models they think will help them best answer these questions. Importantly, they need not worry about whether any model they use to answer one question related to irreversible behaviour is incompatible with any model they use to answer a different question.

6.6 Conclusion

In Sect. 6.4, we showed that the Ehrenfests' statistical mechanics toy models, i.e. the urn model (Sect. 6.3.1) and the P–Q model (Sect. 6.3.2), fulfil the function usually ascribed to toy models (Sect. 6.2): they show that an irreversible entropy evolution and time-symmetric dynamics are compatible properties of statistical mechanical system and demonstrate that such irreversibility is an implication of the Stoßzahlansatz. As such, the Ehrenfests' toy models provide a further illustration of the investigative functions often performed toy models, as identified by Luczak (2017).

Furthermore, we demonstrated (Sect. 6.5) that the non-representative nature of toy models allows a group of such models to be used in a complementary way: different toy models can be designed to answer different questions about a system without any constraints on their compatibility. Accordingly, rather than being a disadvantage or defect, their lack of representation makes them uniquely effective as investigative tools.

The Ehrenfests clearly realised the investigative advantages of toy models. Their introduction and use of the urn and P–Q model was clearly instrumental in popularising the use of toy model in statistical mechanics and prepared the ground for the development of the even more influential Kac-ring model.

References

Alexander, R., Dominik, H., & Stephan, H. (Forthcoming). Understanding (with) toy models. *The British Journal for the Philosophy of Science*.

Boltzmann L. (1872). Weitere Studien über das Wärmegleichgewicht unter Gasmolekülen. *Wiener Berichte*, *66*, 275–370.

Brown, H. R., Myrvold, W., & Uffink, J. (2009). Boltzmann's H-theorem, its discontents, and the birth of statistical mechanics. *Studies in History and Philosophy of Science Part B*, *40*(2), 174–191. ISSN 1355-2198. https://doi.org/10.1016/j.shpsb.2009.03.003, http://www.sciencedirect.com/science/article/pii/S1355219809000124.

Ehrenfest, P., & Ehrenfest, T. (1907). Über zwei bekannte Einwande gegen das Boltzmannsche H-Theorem. *Physikalische Zeitschrift*, *8*, 311–314.

Ehrenfest, P., & Ehrenfest, T. (1909). *Begriffliche Grundlagen der statistischen Auffassung in der Mechanik*. Leibzig: Teuber.

Frigg, R., & Nguyen, J. (2016). Scientific representation. In E. N. Zalta (Ed.), *The stanford encyclopedia of philosophy*. Metaphysics Research Lab, Stanford University, winter 2016 edition.

Frigg, R., & Nguyen, J. (2017). Models and representation. In L. Magnani & T. Bertolotti (Eds.), *Springer handbook of model-based science*. Berlin and New York: Springer.

James, N. (2019). It's Not a Game: Accurate Representation with Toy Models. *The British Journal for the Philosophy of Science*.

Kac, M. (1956). Some remarks on the use of probability in classical statistical mechanics. *Bulletin of the Royal Belgium Academy of Science*, *53*, 356–361.

Luczak, J. (2016). On how to approach the approach to equilibrium. *Philosophy of Science*, *83*(3), 393–411.

Luczak, J. (2017). Talk about toy models. *Studies in History and Philosophy of Modern Physics*, *57*, 1–7.

Paul B. (2013). Analogy and analogical reasoning. In E. N. Zalta (Ed.), *The stanford encyclopedia of philosophy*. Fall 2013 edition.

Uffink, J. (2007). Compendium of the foundations of classical statistical physics. In J. Butterfield, & J. Earman (Eds.), *Handbook for the philosophy of physics* (pp. 924–1074). Amsterdam: Elsevier. http://philsci-archive.pitt.edu/2691/.

Part III
Translations From German and Dutch

Chapter 7
Translation from German: Foundations of Thermodynamics 1925 and 1956

Marina Baldissera Pacchetti

Afanassjewa Foundations of Thermodynamics 1956

Preface (VORWORT) pp. IX–XII

[IX] The most important motive that stimulated the research on the laws that make up the subject of thermodynamics was the wish to find the cheapest way to obtain work from the forces of nature.[1]

This motive also prompted the popular formulations of the different laws:

I: "It is not possible to obtain work from nothing" (i.e., without consuming a corresponding quantity of work from nature);

II: "It is not possible to let a periodic machine turn heat into work by using only one single heat reservoir of a particular temperature" (i.e., one always has to release part of the obtained heat into the environment [which is kept] at a lower temperature);

III: "It is not possible to reach the absolute zeroth value of the temperature" (in fact, one can see that at this temperature the quantity of heat released by a cyclic process would be zero).

As one can see, all of these are answers to the different aspects of one and the same question regarding the most economical use of the forces of nature in order to produce the work desired by mankind—and they are pessimistic answers!

Since its conception, thermodynamics was weighted down with worries about economy [of heat conversion], and these worries have had not only a stimulating but also, in hindsight, a hindering influence on its development—as they do in the developments of the human individual.

This hindrance affected the correct understanding of the second law and those theories of the structure of matter that attempt to explain it.

[1] P. Epstein has given evidence of interesting data in his *Textbook of Thermodynamics*, which shows the important contribution to the discovery of the first law by the Medici.

M. Baldissera Pacchetti (✉)
School of Earth and Environment, University of Leeds, Leeds LS2 9JT, UK
e-mail: m.baldisserapacchetti@leeds.ac.uk

J. Uffink et al. (eds.), *The Legacy of Tatjana Afanassjewa*, Women in the History of Philosophy and Sciences 7, https://doi.org/10.1007/978-3-030-47971-8_7

[X] It was natural for the engineer, from whose hands the physicist received the problem of thermodynamics, to see all the causes that decreased the "efficiency coefficient" as one general evil, and not to divide them into categories. So it happened that in almost[2] all textbooks, from the beginning until now, the representation of the second law that is closely connected with the "economic coefficients," has been tied together with the dissipation of energy and with the one-sidedness of the processes of nature, despite the fact that this dissipation is relevant only for the economic coefficients but not for the actual nature of the second Law—namely for the fact that the expression $\frac{\Delta Q}{T}$ is a complete differential.

This viewpoint is not only harmful in so far as it unnecessarily complicates one of the fundamental formulas of thermodynamics and makes its derivation hard to understand, but also because it diverted the attention of the physicist onto the wrong tracks at a particular point (namely at the point of handling the Boltzmannian theory of the H-Function[3]), and it was the cause of an endless flood of publications that strived to reconciliate logically incompatible things with one another.

If I dare to produce one more book among so many, and in some cases excellent, textbooks on the same topic, then it is to show how one can free the derivation of the fundamental thermodynamic equations from the question regarding the direction of natural phenomena and from the dissipation of energy.

On the one hand, in addition to this, it appeared desirable to me to deal one more time thoroughly with a series of concepts that usually are taken as clear, without really being clear. On the other hand, I have here given an account of my attempt [XI] to more detailedly treat the so-called "irreversible processes" in terms of pure thermodynamics (i.e., independently of any hypothesis concerning the underlying structure of matter), and at least to outline a proof for the increase of entropy (although not without the explicit introduction of a very plausible hypothesis—Hyp. III, 3). This is something that, as far as I know, has not been carried out by anyone until now (one usually does not go further than to show that the entropy cannot decrease).

The separation of the second law from the question regarding the dissipation of energy makes it possible to illuminate more clearly the relation between the theory of the H-Function and thermodynamics. One can show that the one-sidedness of the trajectory of the phenomena, that Boltzmann wanted to prove—but could not—is irrelevant for the validity of the equations of thermodynamics, and that independently of this, the H-Function is of great importance for the explanation of those equations, that make up the actual content of the second law.[4]

This book is therefore dedicated to the representation of principles. Accordingly, it does not contain any exact descriptions of real experiments nor any applications;

[2]The third edition of the textbook by V. D. Waals, edited by prof. Ph. Kohnstamm, does contain a treatment of the "Second Law" (Zw.H) directly related to the receptions of this book. This is also partly the case for the course by L. G. de Haas-Lorenz. Compare also with A. Landé: *Axiomatiske Begründung der Thermodynamik von* C. Caratheodory. Handb. d. Phys. Bd. IX, p.282, §I. (Herausgeg. C. H. Geiger u. K. Scheel Verl. Julius Springer.)

[3]Regarding the true meaning of the H-Function for the "Second Law," c.f. appendix II, and regarding the reconciliation incompatible things, c.f. appendix III.

[4]C.f. citation.

it only contains thought experiments and very simple illustrations in order to clarify one concept or another. This book also does not aim to do justice to the great founders of thermodynamics by providing complete historical quotations. These two aspects of the subject can be found in so many existing courses, and its incorporation into the actual theme of this book would make the argument of this book more difficult.

Nevertheless, I hope that it can also be used by the beginner as a useful addition to the usual course—at least as long as the approach that I have imagined has not yet become part of the didactical practice. Above all, however, this book is directed at lecturers of thermodynamics: it is my experience that a careful discussion of the foundational concepts makes the understanding of the material easier.

[XII] A mathematical chapter is appended to the actual course of Thermodynamics,[5] dedicated to the theory of the so-called Pfaffian equations (Pfaff'schen Gleichungen), in so far as these are necessary for the understanding of thermodynamics. This theory is not taught in universities as a mandatory topic for all physics students, so that these students have to deal simultaneously with two difficulties when they are taught thermodynamics. This is what would happen to someone, if they had to learn about the fluctuations of currents in their galvanic chain without having first learned about the cosine function. As an aside, this theory will not be necessary before chapter IV.

Afanassjewa Foundations of Thermodynamics 1956

Chapter I pp. 1–16

[1]

First Chapter

Introductory Remarks

§ 1. Parameters of State

We will call “a system” any combination of bodies whose properties are studied in thermodynamics. In special cases, the system also may be a single homogeneous body. We distinguish between the “equilibrium” and the “out of equilibrium” states of a system. A system is in “equilibrium” if a state is maintained permanently, except if there are influences from the environment that can disturb this equilibrium. A system is “out of equilibrium” if it cannot maintain a state permanently: this inability is itself the cause of its [the state's] changes.

Whether a system is in equilibrium, or not, depends on the distribution of particular quantities, which characterize its state and which we will call “parameters of state,” over its different parts and on the structure of the system (i.e., on the connection between the different parts of the system and its relation to the environment).[6]

The equilibrium state is completely determined by the values of a fixed number n of its parameters of state. This can be clarified by some easy examples.

[5] Appendix I, Note 2, p. X

[6] Our definition of “equilibrium” relies, indeed, on an idealized extrapolation of experience: nowadays one also thinks about the spontaneous departure from a state that remains unchanged for and an extremely long period of time: this in connection with the meaning of the thermodynamics quantities by means of the kinetic theory. The quantum-mechanical interpretation also does not assume any infinitely lasting states of equilibrium.

Example 1. The state of equilibrium of a quantity of a gas in a cylinder sealed by a piston can be described unambiguously, by giving the value of its volume and its pressure on the piston [2] because one can determine its temperature, density, etc. from these. This is a system for which n = 2.

Example 2. Two quantities of gas that are in a cylinder closed on both sides by moving pistons and that are themselves separated by a third piston, can be made into a system with n = 3 in various ways:

2a. The middle piston is unmovable but allows for exchange of heat. In this case, the equilibrium state is determined by a common temperature τ and by the pressures p_1 and p_2.

2b. The middle piston can move freely but it does not allow for the exchange of heat. In this case, the equilibrium is only possible when the pressures of the two parts have the same value p. The temperatures τ_1 and τ_2, however, can be indefinitely different. The volumes, etc., are then determined by the three quantities p, τ_1, τ_2.

If the middle piston is both movable and allows for the exchange of heat, then we obtain a system with n = 2.

The equilibrium states described above can, however, only be maintained if the conditions of the environment are adjusted to allow this: either one should require a complete isolation from its environment for all three examples, or all bodies that are in contact with the different parts of the system should have the same temperature as those parts. The pressure on the outer side of the piston should also be the same as the pressure exerted on the piston by the corresponding internal parts of the system.

In order to describe the equilibrium state of the system, one can choose any other parameters instead of the ones just used, as long as the former are functions of the latter: one can calculate the value of the ones from the equations that relate the new parameters with the others, if one knows the latter (obviously only if the solutions to the equations are unique!).

It is only essential that an equilibrium state of a system—depending on its structure—is always determined by a certain number of independent parameters. This means the values of all other parameters are completely determined by these.

[3]

§ 2. Point parameters and additive parameters

We now want to highlight certain types of parameters of state.

"Point parameter": one can describe as such those parameters whose values are not tied to the extensions and masses of the matter they describe. This is to be understood in the following way: if one were to virtually decompose the system into much smaller subparts, then it would be possible to ascribe the same value of these parameters of the system to each of these subparts. In this way, it makes sense—if one thinks about making these parts indefinitely small—to assign the corresponding value of the parameter to each point of the system (which, indeed, is an idealization, as one does not, nowadays, think of matter as a continuum. However, this idealization is often useful for that category of problems to which thermodynamics is primarily applied).

Examples of point parameters: temperature, pressure per unit area, density, dielectric constant, etc., and, obviously, all functions of these parameters.

"Additive Parameters": one value of such a parameter is given to each system, as well as to each of its parts, in any given state. However, the value of the parameter given to the whole system is the sum of the values of the parameter of all the parts. For this parameter, one can also use the name "quantity parameter" or "extension parameter."

We call a system homogeneous if all of its point parameters have the same value at each point. This way we can say that the values of the additive parameters of a homogeneous system are proportional to one another.

Examples of additive parameters are mass, volume, weight, and in many cases energy[7] (in contrast to the "specific" quantities: mass per unit volume or per Mol., which belong to the point parameters).

One can say that the point parameters characterize the state of the matter that makes up the system, while the additive parameters are additionally determined by the structure of the system.

[4]

§ 3. Conditions for equilibrium

The structure of the system can be such that the values of a particular point parameter of the different parts of the system can be arbitrarily different from one another, without having any influence on the equilibrium of the system. In this case, we say that these parts are "*isolated*" with respect to this parameter. If an "adiabatic" wall—a wall that does not allow for the transfer of heat—divides two parts [of the system], then these two parts are isolated from one another with respect to temperature ("*thermically*").

If both parts are gaseous and are separated by an unmovable wall (such as a fixed piston), then they are isolated from one another with respect to pressure ("*mechanically*").

If, however, the structure of the system is such that is it necessary for the maintenance of its equilibrium that the values of the two parts of the system are in a particular relation to each other—"*Condition for equilibrium*"— then we say that these parts are "*coupled*" to one another "*with respect to this parameter.*"

Two parts that are in direct contact are clearly coupled with respect to all parameters that have an influence on equilibrium.

The conditions for equilibrium have different forms for different parameters. As such, the condition for equilibrium for a thermal coupling is

$$t_1 = t_2,$$

where τ_1 and τ_2 are the temperatures of the two parts, respectively. For the mechanical coupling, in the case of direct contact, is $p_1 = p_2$, where p_1 and p_2 are the pressure per unit area in the two parts, respectively.

[7]C.f. §25.

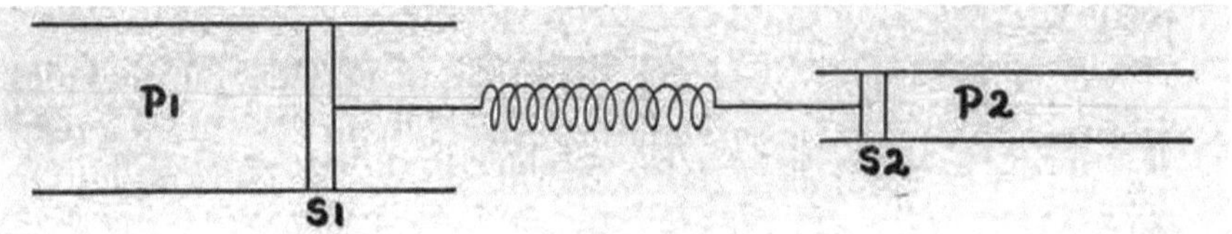

Fig. 1

[5]
However, this can be different for the case of an indirect mechanical coupling. For example, when two gases do work on one another by mean of two differently large pistons, that push on opposite sides of a spring (Fig. 1), then the condition for equilibrium is

$$F_1 = F_2,$$

where $F_1 = F_2$ are the forces that both systems, respectively, exert on the spring. It follows that

$$p_1 S_1 = p_2 S_2.$$

Here, S_1 and S_2 are the areas of the pistons.

If the system is made of two different metals that touch one another, then the condition for equilibrium for the "*electrostatic*" coupling is

$$V_2 = V_1 + a$$

where V_2 and V_1 are, respectively, the values of the electric potentials [of the two metals] and a is the change in potential at the surface of contact of the two metals.

If two parts of the system that are in contact are made of one and the same material but are in two different phase-states, then the condition for which no matter is exchanged between one part to the other is, as we will see later (Ch. IX, § 59), given by the equation:

$$u_1 - Ts_1 + pv_1 = u_2 - Ts_2 + pv_2,$$

where all letters represent point parameters.

§ 4. Contact and Distance couplings

In the previous paragraphs, we have only focused on "*contact couplings,*" in which the two parts of the system are either in direct contact or are in contact by means of other bodies. There are also other kinds of couplings that can destroy the equilibrium: the "*distance couplings.*"

For example, a stone is connected to the earth by a "*distance coupling,*" if there is no obstacle to its fall to the ground under the influence of the force of gravity. Similarly, two electrical charges on two parts of a piece of metal are coupled distantly.

[6]

Let us compare the mechanic contact coupling—the pressure of two gases on a movable piston that divides the gases—with the distance coupling: the mutual attraction of two gravitating bodies. If the pressures of the gases are the same, then there is equilibrium, because both parts exert the same force in two opposite directions on the same body—the piston. In the case of two gravitating masses that are at a distance from each other, they each suffer from the effect of gravity only in one direction and therefore must move (if no other obstacles are present).

§ 6. Spontaneous processes

A "*process*" starts if an equilibrium is destroyed. A process is a chronological sequence of always new non-equilibrium states. Under certain circumstances, a new state of equilibrium can be attained, and in this case, the process comes to an end. N.B. In principle, this does not always need to be the case, as this example shows: two quantities of gas, that partially fill a tube, are separated from one another by a piston S and from the environment by pistons S_1 and S_2. Two constant and unequal pressures p_1 and p_2 ($p_1 > p_2$) act on the pistons S_1 and S_2. If S is screwed on tight, then the system can be in equilibrium. If one lets S be movable, then it will move in the direction of the larger to the lesser pressure; at the same time, however, the outer pistons will also move in the same direction, and the process will end only with the destruction of the original system—if the piston S_2 shoots out of the tube: a new equilibrium state of the given system is impossible!

The cause of the disruption of the equilibrium is the removal of the separation—and the setting up of a new coupling between two parts of the system under the absence of the corresponding conditions for equilibrium (or the setting up of a coupling between the system and the environment, which corresponds to the same thing, as the coupling now concerns a larger system).

To analyze the process, we can mentally divide each originally homogeneous part of the system into a very large number of smaller parts [7]. The corresponding point parameters will all be the same for all subparts of the system only for one instant after the establishment of the appropriate kind of coupling: every value of the parameter for each subpart that is located near the coupling surface will change immediately afterwards (and in particular, the change will mostly occur in such a direction as to balance the values of the parameters on both sides of this surface). This creates an inequality between the parameters that lie on this surface and the layers on the other side of the surface, which leads to a further change in the values of the parameters. In this way, the disruption of the equilibrium state penetrates further across the subparts of the system. If, however, a particular parameter in one of the subparts is changed, then this causes a change of many other parameters that are functionally tied to the original one. In addition, for fluids and in particular for gases, the difference in pressure on the two sides sets each subpart in motion.

This way we obtain a state of non-equilibrium that has to be described by an enormous number of values of the equilibrium parameters, and eventually by a correspondingly enormous amount of components of velocity. How large the number of subparts the system should be divided in is, naturally, unspecified: our description can only be approximate; so the possible accuracy is limited by the fact that it is known that matter is not seen as a continuum, such that the ongoing division into subparts would lead past a particular boundary to objects for which terms like "pressure" and "temperature," etc. would not be applicable.

If, however, we are satisfied by the approximate description sketched above, then we can obtain a deeper insight into the nature of the thermodynamic phenomena and better assess the limits of validity of the different laws (cf. Ch. VI, §§ 42–44).

In any case, we may not forget, that the domain of applicability of this kind of description does not apply to all possible non-equilibrium states: for the case of very turbulent changes of state, in which two streams of molecules move chaotically in different directions in one and the same area, the terms "temperature" and "pressure" lose their meaning even for not very small subparts.

§ 6. Graphic representation

It is useful to summarize the above by means of a graphic representation. We want to illustrate the use of this method with the easy example 2, §1 of this chapter. We assume, that the middle piston is diathermal.

Let us imagine that the system is in equilibrium. The common temperature of both parts [of the system] is τ^0; and the pressures are $p_1^0 > p_2^0$; the volume of the first part is v_1^0, and of the second part v_2^0.

Since there is a relationship (equation of state) between temperature, pressure, and volume of a homogeneous gas, any equilibrium state is totally determined if the value of any one of the three parameters (e.g., τ^0, p_1^0, p_2^0) is given.

We now want to represent these three independent parameters as perpendicular coordinates of a three-dimensional space R_3. Imagine that when such a state occurs, both external pistons stay fixed, but the middle piston is movable.

In the first instant, the state can still be represented by the same point in R_3, but immediately afterwards there is a change; in addition to the thermal coupling, a mechanical coupling is created between the two parts, and an equilibrium is possible only when $p_1 = p_2$. The middle piston now is under the influence of the force

$$F = \left(p_1^0 - p_2^0\right)S,$$

where S is the surface [of the piston]; it will start moving from the higher to the lower pressure. This causes a compression and a dilation, respectively, in the very thin layers [of gas] lying directly next to the piston. The pressures in these layers change accordingly, and it is not the same as the respective pressures of the rest of the two parts of the system. This causes changes in the densities of the consecutive layers [of the gas], but it also causes changes in temperatures and the creation of more or less significant velocities of each singular layer of gas taken as a whole (one sees, how the approximative character of our description resurfaces).

[9]

Next, we want to observe a first, short interval of time of the process. The velocities of the piston and of all the layers of gas are negligible in this case.

How can the state at every moment of this interval of time be represented graphically?

If we mentally divide out the system in $N = N_1 + N_2$ layers, so that we can assign a value τ_2 and a value p_2 to the i-th layer, then we only need a 2 N-dimensional space R_{2N} for the graphic representation. Every instantaneous state of the system is represented by a point and the succession of non-equilibrium states is represented by a line—a "*path*"—in R_{2N}.

The original R_3 is represented as a subspace of this R_{2N}, and in particular as a "*Diagonal space,*" if it is allowed to call each subspace such, which fulfills a particular number of equations of the type

$$p_k = p_1 = \ldots; \tau_m = \tau_n = \ldots$$

We, therefore, have, after the establishment of the mechanical coupling that disturbs the equilibrium of the system, the representational point leaves the three-dimensional diagonal space, which fulfills the following equations:

$$\tau_i = \tau(i = 1, 2, \ldots N) \qquad (a)$$

$$p_i = p_1(i = 1, 2, \ldots N_1) \qquad (b)$$

$$p_i = p_2(i = N_1 + 1, N_1 + 2, \ldots N_1 + N_2) \qquad (c)$$

The system was in the three-dimensional diagonal space in the first instant of the process.

If the process is such that in the subsequent stages the velocity of all the layers is also negligible, then the space R_{2N} is sufficient for the representation of the whole process: the representational point then represents a path within this space.

We know that the process—whatever its intermediate states may be—will have, given the described arrangement of the system, an end. Its last instant is given by a new equilibrium state, and, in particular, by such a state that fulfills the following equation:

$$p_1 = p_2$$

in addition to equations (a), (b), and (c). [10] The endpoint of our path is in the diagonal space R_2 of the subspace R_3, which corresponds to the fact that the state of equilibrium is determined by a small number of independent parameters for the new structure of the system that was created by the new coupling.

If the velocities are significant throughout the process, then they have to be added as parameters of state to the original ones, and the representational space should correspondingly have more dimensions.

It is clear how the general representation will look: if, at equilibrium, the system is composed of many homogeneous parts and if the equilibrium is disturbed by the fact that new couplings are created between k of the n independent parameters, whereby the equality of those parameters is necessary for the equilibrium, then the total process will be represented by a path in R_N if it ends with a new state of equilibrium. The start of the path will be in the diagonal space R_n and the endpoint of the path will be in the diagonal space R_{n-k}.

§ 7. Infinitely slow, spontaneous processes

We now want to consider an idealized, extreme class of processes, i.e., the "*infinitely slow*" processes. We will restrict ourselves only to those cases, for which the establishment of a new coupling between two spatially separated parts (*phases*) of the system is the cause of the process.

The investigation of such processes is relevant because it can prevent us from confusing them with the so-called "*reversible*" processes, which also occur "infinitely slowly."

We will, once again, make use of example 2, § 1 of this chapter. Our system will be both thermally and mechanically isolated from its environment. The latter implies that both external pistons are immovable. The initial state is the same as the one assumed in the above paragraph.

The process can be slowed down in different ways:

A. The frictional resistance during the movement of the piston can be arbitrarily large. In this case, it will move arbitrarily slowly to the position that corresponds to equal pressures [of the two parts of the system].

[11]

B. The friction can be negligible, but the slowing down is obtained by stopping the middle piston arbitrarily often and for the corresponding equilibrium to be reached: in this case, no fast velocities will develop at any stage, and the sum of the intervals in which the piston will have to move to its final position can be made arbitrarily large by having a large enough number of interruptions [to the movement of the piston].

Obviously, there is no perfect infinitely slow process. Therefore, we cannot say that the just described arbitrarily slow processes have an infinitely slow process as a limiting case. But the properties of the non-equilibrium states that the system goes through approach a continuous sequence of equilibrium states, which connect the initial to the final state. When we use the term "*infinitely slow*" for the processes described above, then we mean that the processes in question are sufficiently slow, so that we can ignore particular deviations from the equilibrium states for each of its states. For example, at some point in the previous paragraph, we have ignored the velocities; we could have summarized this in the following words: "if we have an infinitely slow process, then a R_{2N} is sufficient for its graphical representation, since the velocities can be set equal to zero."

The equilibration of temperature between two parts [of a system] can also be slowed down analogously.

However, we cannot ignore the following: however close every instantaneous state can be to an equilibrium state during this infinitely slow process, the cause of the process is the deviation of its parameter values from the values at equilibrium, i.e., the leap of the values at the boundary between the two parts [of the system]—something that belongs to the structure of the system and each of its current instantaneous states.

Therefore, we will call all these kind of processes (slow ones but also not slow ones) "*spontaneous.*"

[12]

§ 8. Forced changes of state

We now want to contrast spontaneous processes and "*forced changes of state,*" which are caused by the disturbance of equilibrium due to a coupling of the system with its environment.

The nature of the cause will be always the same: jumps in the value of any of the intensity parameter, that violate the conditions of equilibrium. However, the number of independent parameters that determine the newly established state of equilibrium of the given system, remains, in general, the same as the one before the coupling: such a process has definitely run its course when the values of the relevant parameters of the given system and of the environment are equilibrated, not those [parameters] of two parts of the system itself. Moreover, one can stop such a process by breaking the coupling sooner.

While, for a spontaneous process, one assumes a change in the structure of the given system (a new coupling between its parts), one can, with the help of forced changes of state, let the system assume all possible states of equilibrium that are compatible with one and the same structure.

A particular kind of forced changes of state are particularly important for (theoretical) thermodynamic investigations: those for which the values of the parameters of the given system and of the environment, which is coupled through these parameters, almost correspond to the equilibrium conditions. Those have the following characteristics: the difference between the initial and final values of the parameters of the given system are arbitrarily small; the state of the system throughout the whole process is virtually a state of equilibrium; the values of the intensity parameter, with which the system is coupled to the environment, can be equated with those of the environment without incurring in notable errors; the average velocity of the equilibration of the parameters – $\frac{q^1-q}{\Delta t}$ – where q is the (common, of both systems[)] initial value and q^1 is the final value of the relevant parameters, and Δt is the duration of the process, is—in general—for the different kinds of parameters, smaller, the smaller the difference $q - \bar{q}$ between the two systems is at the beginning of the process and also the smaller the difference $q^1 - q$ is (as q^1 is between the values q and $\bar{q}$) [13]. The graphical representation of such changes of state is a part of a curve in R_N, both ends of which lie in R_n in the subspace of the equilibrium parameters of the given system, and whose remaining points lie very close to this R_n.

We will call these "*elementary quasi-static processes.*"

§ 9. Quasi-processes and quasi-static processes

In the following, we will occupy ourselves a lot with continuous sequences of equilibrium states that join two given states of equilibrium. They will be represented graphically by curves in R_n. Traditionally, they have been called "processes," and in particular "reversible" processes. We will rather call them "*Quasi-processes,*" as—clearly—they cannot be actualized by any real process; further, we want to explicitly leave out the epithet "reversible" (see § 10).

Despite the fact that strictly a quasi-process cannot be actualized, one can nevertheless let the system assume an arbitrarily dense discrete succession of equilibrium states that belong to the system, in such a way that, even throughout processes that connect two consecutive equilibrium states, the state of the system is almost in equilibrium: [this is achieved] by letting the system go through a sequence of elementary quasi-static processes. We call such a sequence "*quasi-static process*" (cf. C. CARATHEODORY, *Untersuchungen über die Grundlagen der Thermodynamik (Math. Ann.* 67 [1909], 355)).

Because the velocity of an elementary quasi-static process tends to zero if the differences between the parameters of its beginning and end points are shrunk, the duration of the whole quasi-static process tends to infinity as the number of its elements increases. A quasi-static process is therefore also an "*infinitely slow*" process. However, the essential difference between a quasi-static process and an infinitely slow spontaneous process should not be forgotten:

A. A quasi-static process is a forced process.

[14]

B. It is not a simple process, but a sequence of processes, for the realization of which the system should be consecutively coupled to a sequence of external systems. (Perhaps it might be coupled to a single system, but in this case, a change has to occur to this system at each step, to adjust its state to the given system).

For later use, we will introduce the following terms: A spontaneous process, whose initial and final state are equilibrium states that will be called a "complete" process. A quasi-process, whose initial and end states are identical with the ones of a complete spontaneous process, will be called an "*equivalent process*" to one of the latter [processes described].

N.B. Obviously, one can assign to a given complete process infinitely many equivalent quasi-processes. In the case in which the spontaneous process is infinitesimally small (i.e., if the changes in the parameter x_i are expressed by the differentials dx_i), then the infinitesimal equivalent process is uniquely determined—among all other possible equivalent processes.

§ 10. "Reversible" and "Irreversible" processes

The notions of "reversible" and "irreversible" processes have played a prominent role in the history of thermodynamics. Their use, however, is not unambiguous. We, therefore, need to dwell on this a little.

We will call a process an "*exact inverse*" to a given process, if the system goes through all the steps of the given process in an exactly reversed sequence (with reversed velocity of the subparts, obviously).

We will call a process "*reversible,*" if its exact inverse process can also occur in nature. Otherwise, we call [this process] "*irreversible.*" Since our everyday experiences teach us that spontaneous processes only have one direction, these will also be called “irreversible.” It should be noted, however, that the interpretation of the thermodynamic phenomena through the modern theories of the structure of matter makes it conceivable, that, in time, such time periods should occur, during which the course of the phenomena [15] should be (almost) inverse. Therefore, the “irreversibility” of spontaneous processes is better understood in the sense that these can only have one direction in our lifetime.

We have maintained this sense of irreversibility also in the previous arguments: we have always considered it self-evident, that the intensity parameters equilibrate themselves, if they are given an opportunity through the establishment of a coupling; we did not think that after such an equilibration the temperature of both parts would change spontaneously, and that at the end of this disintegrating process, these parts would have two different temperatures from the ones they had at the beginning of the previous process! We will always acknowledge the fact that motion will be dampened by friction or by the collision of non-elastic bodies and heat is created in the process, and that in our lifetime, no corresponding exact (or nearly exact) inverse process could be observed.

However, in any case, the name “irreversible process” does not have the unequivocal meaning that it initially had.

This was only one of the meaning of this term: one uses the “reversible–irreversible” contrast also to contrast real processes with what we have called here “quasi-processes.”

Since the possibility was acknowledged, that real processes may not have one and the same direction for all eternity, the term “reversible” cannot be used anymore to distinguish quasi-process from real processes.

There are other reasons why the choice of this distinction was not a fortunate one: quasi-processes are not processes, so there is nothing [of this processes] that can be reversed—unless, one wants to say that one can let them [quasi-static processes] run virtually in both directions. However, this is also not forbidden for the sequence of non-equilibrium states that a system runs through during a real process. If one does not image an exact quasi-process, but an approximating quasi-static process, then these are reversible in the same sense as all other real processes are. Correspondingly, a quasi-static process that approximates a quasi-process is not exactly the inverse of [the process] that approximates the reversed quasi-process: [16] for example, if the given quasi-process constitutes a continuous flux of heat to the given system, then the systems of the environment that are used [for this process], will have higher temperature than the given system for the original direction [of the process], and lower temperatures than the given system for the reversed direction [of the process].

Finally, one can talk about adiabatic “reversible or irreversible” processes in a very peculiar way: if an adiabatic process leads a system from an equilibrium state A to an equilibrium state B, and if there is another also adiabatic process, that can link the end state B to the initial state A, then one calls the initial adiabatic process “reversible.” If the initial change of state is a quasi-process, then it is reversible by definition. Therefore, the question of reversibility only has meaning when it concerns a real adiabatic process.

For the following (ch. IV), it is very important to determine whether the possible adiabatic changes of state from the end state B to the initial state A is a quasi-process or a real process. We will announce here already, that the only important thing for the derivation of the thermodynamic equations is the axiom that can be described in the following words:

"When a system is taken by a *real* adiabatic process from A to B, then it cannot be taken by any adiabatic *quasi-process* from B to A." (Whether the initial real adiabatic process is reversible, in the first here described sense of "reversible," does not have any influence on the conclusions regarding the derivation of the thermodynamic equations.)

[57]

Sixth Chapter

Real Processes

§ 38. Completed real processes and equivalent quasi-processes

In the previous analyses we have talked mostly about quasi-processes and we have set up relations that determine the properties of systems in states of equilibrium. These properties have been the main topic of thermodynamic investigations (until recent times at least), too. In order to learn anything from them, however, one has to observe phenomena, i.e., real processes. Real processes, with all the properties that distinguish them from quasi-processes, are also the objects with which technicians concern themselves. In order to make a quasi-process accessible to experimental investigation or in order to connect it virtually to thought experiments, we have to assume that they are approximated by quasi-static processes. These are real, albeit idealized, processes.

Previously we could therefore not leave real processes completely unmentioned. For example, the difference between "equal" and "unequal temperatures" is based on the possibility of real processes. At one point in Chapter III, which deals with the concept of "energy," we could not go on without considering an explicit non-quasi-static process: [that point] when we were dealing with the concept of "quantity of heat" (§19).

Furthermore, some questions about real processes will become clearer if, in addition, we observe certain quasi-processes. This is the case for the question about the direction of real processes and about the change in entropy of adiabatic real processes.

The direction of a process can be characterized by stating the changes in energy and entropy that accompany it. In order to be able to state these clearly, [58] one has to express the quantities that change the different forms of energy as functions of the equilibrium parameters, and the changes thereof, that belong to the starting state and the end state of the process; assuming that the process in question is a completed process. In this case, then, the changes in energy, and also in entropy, are the same as in any equivalent quasi-process.

§ 39. Hypotheses of the third group

In Chapter II, §2, we already talked about a special kind of real processes—about the transformation of energy into heat by mechanical contact coupling. Here, we want to establish, through posing specific hypotheses, what can be assumed in the most general cases about the trajectory of real processes.

Hypothesis III, 1: *for a spontaneous process to take place, it is necessary, that at least one equilibrium condition is broken (is not met anymore): that either particular couplings are created between the parts of the system, where, in the previous state of equilibrium, the corresponding parameters had values that do not suffice for the new conditions of equilibrium; or that certain parts [of the system], among which there are chemical affinities, are brought in direct contact; or that devices, that counteract particular forces [that act at a distance (Fernkräften)], are removed.*

Hypothesis III, 2: *if conditions for the destruction of the equilibrium obtain, then a spontaneous process actually begins.*

Hypothesis III, 3a: *a spontaneous process always occurs in such a way that, between the two coupled parts of the system, heat is transferred from the part with the higher temperature to the part with the lower temperature.*

Hypothesis III, 3b: *if, through a spontaneous process in an isolated system, the energy of a particular kind A in one part [of the system] is decreased by the amount* $|DU_A^1|$, *and in another coupled part is increased by the amount* $|DU_A^2|$; *or if energy of kind B is decreased by an amount* $|DU_B|$, *and energy of kind C is increased by the amount* $|DU_C|$, *then it is always the case that*

$$\left.\begin{aligned} |DU_A^1| &= |DU_A^2| + \bar{Q} \qquad \bar{Q} > 0 \\ |DU_B| &= |DU_C| + \bar{Q} \qquad \bar{Q} > 0 \end{aligned}\right\} \tag{17}$$

[59] *where, in both cases,* $\bar{Q}$ *represents the quantity of heat that is created in the system at the cost of* $|DU_A^1|$ *and* $|DU_B|$, *respectively.* This also means: if

$$\left.\begin{aligned} &DU_A^1 < 0, DU_A^2 > 0, \text{ then } \bar{Q} > 0 \\ &DU_B < 0, DU_C > 0, \text{ then } \bar{Q} > 0 \end{aligned}\right\} \tag{18}$$

Naturally, the kinds [of energy] A, B, C are meant to be different from heat.

The equations (17) are to be understood in the following sense,—if one thinks of these energy transfers to be isolated from the others — then there always exists an equivalent quasi-process, which contains a positive heat transfer as one of its constituting parts. It follows from hypothesis II, 2, that the heat transfer will also be positive for infinitely many other equivalent quasi-processes: if the heat transfer for a particular quasi-process is positive, then the change in entropy is also positive; however, since the entropy is represented as a unique function of the equilibrium parameters of the system, the change in entropy will be the same for all equivalent quasi-processes.

The hypotheses that we have just established imply that, of two processes that are the exact inverse of one another (see Chapter I, §10), only one is possible, namely the one for which different forms of non-heat energy [von Wärme verschiedenen

Energieformen] partly become heat, and for which heat, in the case of thermal coupling, is transferred from the warmer to the colder part [of the system].

Therefore, as a consequence of the hypotheses of the third group, real processes are "irreversible" in the sense that their inverse is impossible.

§ 40. Example

We want to explain the meaning of the equivalent quasi-processes for the calculation of the energy conversion on the well-known spontaneous processes of pressure-equalization. The setup is hence given by example 2, Ch. I, §1, whereby, for simplicity, we assume that the middle piston is diathermal, and therefore $T_1 = T_2$. The initial pressures are unequal- $p_1 > p_2$. The middle piston is released for a short amount of time only, after which it is fixed again. The real process that we observe, is therefore an elementary, complete process. The equally elementary [60] equivalent quasi-process would be the following: one fixes the middle piston, but let the two outside pistons be movable, and through this alteration one couples mechanically each half of our system to the corresponding external system, and, furthermore, one thermally couples the total system, which is thermally homogenous, to its environment. The following equations hold:

$$\Delta Q_1 = dU_1 + \Delta A_1,$$

$$\Delta Q_2 = dU_2 + \Delta A_2,$$

where $\Delta Q = TdS$ and $\Delta A = pd$ v.

Since in our real process the system is assumed to be isolated, we have to take into account that for the equivalent quasi-processes $dU_1 + dU_2 = 0$ is true, where $dv = -dv_1$. Finally, we obtain

$$dU = T(dS_1 + dS_2) - (p_1 - p_2)dv_1 = 0, \text{i.e.,}$$

$$T(dS_1 + dS_2) = TdS = (p_1 - p_2)dv_1.$$

For both processes, the right-hand side of the last equation represents the quasi-potential change in energy of the total system and the left-hand side: [represents] the gain in heat energy. It is, however, instructional to note that the outward effect of both of these forms of energy is different:

For the quasi-process, $(p_1 - p_2)dv_1 = \Delta A$ is the work that the system exerts outward; the expression $TdS = \Delta Q$ is the heat supplied from outside. For the real process both ΔA and ΔQ are equal to zero. Therefore, for the real processes, we have

$$\begin{aligned}(p_1 - p_2)dv_1 \neq \Delta A; TdS \neq \Delta Q; -p_2dv_1 &= DA_1\\ -p_2dv_2 = +p_2dv_1 &= DA_2\\ TdS &= \Delta Q.\end{aligned}$$

§ 41. The concept of the reverse course of real processes

There is a view, according to which it seems credibly that the direction of real processes that is observable today should not remain the same forever: the kinetic theory of matter, which interprets the laws of thermodynamics as consequences of the deeper laws of molecular movements, provides very satisfactory explanations of the properties of the systems while in states of equilibrium and of the laws of quasi-processes. However, concerning the preference in the direction of real processes, on closer inspection, it becomes clear that [61] one cannot find a satisfactory explanation for it, that it would, in fact, be consistent to accept that once in a while—over enormously long intervals of time—such periods should occur, where the trajectory of any phenomenon should have the reversed direction. The same conclusion will be inevitable for any theory that treats the laws of thermodynamics as static results.

Of course, it would not make sense to categorically declare oneself for or against this possibility, in particular since our laws of thermodynamic are mere extrapolations from observations over a relatively small interval of time, and therefore cannot have any claim of validity for eternity, and, furthermore, the static theories can only claim an approximate validity—despite the services they have provided to the study and discovery of different phenomena. However, it may not be superfluous to clarify the logical relations between the laws of thermodynamics and the static theories. While we do not know what we should think about the universe and its most distant future, we have no excuse not to know what we should think about the theories built and applied by physicists!

At this point, we want to clarify, what we should leave in place and what we should change in the hypotheses that we have so far introduced to account for reversed phenomena.

Imagine that a piece of the world—say, the laboratory—is recorded on film [kinematographisch aufgenommen] (in every detail), and that this recording is then rewound in the opposite direction. We want to discuss how a couple of phenomena would appear in such an inversion.

I. A piece of glass is let go from a hand at a particular height. It accelerates to the ground, breaks into many shards, which fall away from one another, and, finally—due to friction with the floor—lie scattered. The theory of kinematics implies that their kinetic energy is transformed into heat. The inverse process, therefore, begins, without being preceded by a disturbance of the equilibrium, with the inverse transformation of the heat energy scattered on the floor and within the shards of glass into kinetic energy, which [62] causes the shards to move, accelerate toward one another in order to form an intact piece of glass, at which point it [the whole piece of glass], starting with a significant velocity, flies upward in a decelerating motion, where, at the appropriate point, an open hand closes around it.

If, however, the original phenomenon was the following: the piece of glass is thrown upward by the hand and then caught again by the same hand, then the reverse trajectory of the phenomenon—for those parts, where the glass is freely flying in the air—is indistinguishable from its original trajectory: a free body, that is subject to the influence of the gravitational force, and that at a particular point reaches zero velocity, in both cases, will consequently accelerate downward.

2. The equilibrium between two solid bodies, which have different temperatures, is broken by putting them in direct contact with one another. Consequently, the process of temperature adjustment begins, and [this process] ends when all parts of the composite system have attained the same temperature (In this we assume, that the whole system has been put in an adiabatic container).

The inverse process starts with this final state, whereupon the distribution of temperature spontaneously starts to become heterogeneous, and in such a way, that the largest inhomogeneity is created at the point of contact between the two bodies, until the particular moment, at which the temperature in all the parts of one body has reached one value, and the temperature [in all the parts] of the other body has reached another value, i.e., those temperature values that they had at the beginning of the direct process. In this instant precisely the bodies are then separated from one another.

Should this last step of the inverse process not occur, then it does not at all follow from the theory of kinetics that the difference in temperature between these two bodies continues to rise: such a thing would not represent an inversion of any process that has been observed to date! On the contrary, everything in this theory indicates that even on the inverse trajectory [of the process], after this moment, a new process of adjustment should develop—exactly as is the case for the inversion of the throwing and falling motion of free bodies, which after the instant where they [the body] have reached zero velocity, will succumb to the gravitational force and fall downward.

[63] It is a different question whether a more or less precise inversion of current phenomena can offer a case, in which, after the setting up of a partially—homogeneous distribution of temperature, no separation of the parts [of the system] that have different temperatures occurs: after all, we want to think of such an inverse process—in connection with its environment, as a constituent part of the inverse process in a larger part of the world! It is however—due to this theory and due to every other statistical theory of thermodynamic phenomena—definitely possible, that such cases occurred in a particular epoch, during which the trajectory of processes was not a precise inversion of the current ones, and in which, nevertheless, the real process partially had this inverse character.

§ 42. Hypotheses that apply to the inverse trajectories [of phenomena]

We now ask: for epochs with inverse, or predominantly inverse phenomena, what would remain of the hypotheses that we set up [previously]?

Clearly, hypothesis III, I of the third group would lose validity, since most processes would arise spontaneously from a state of equilibrium, without new couplings that would destroy the equilibrium state having been established.

Hypothesis III, 3 should be changed in so far as all the signs in the inequalities are changed.

In the case of precise inversions, Hypothesis III, 2 would not have any objects of application, since the state in which two parts of the system are generally in equilibrium, but are coupled to one another by non-maintainable values of the corresponding parameters, is a state of the process upon which a separation of those parts would follow.

In fact, the known statistical theories do not provide reasons to expect a period of exact inversions [of phenomena]. If, however, we are talking about possible periods of approximately inverse trajectories [of processes], during which also such cases in which after the setting up of equilibrium of each of the coupled parts [of the system] the coupling is not removed occur, then, the same adjustment processes that we observe today can easily follow this momentary state: this scenario [65] is not excluded by the statistical theories, rather, it seems to follow from them. Therefore, we can say that for approximately inverse trajectories [of processes], hypothesis III, 2 would retain its validity … in so far as objects of application for it are available. However, in this case, the change of signs in the inequalities (18) in Hyp. III, 3 would not always, but only often, be necessary: hypothesis III would, therefore, lose validity.

The hypotheses of the second group, which exclusively concern equilibrium states, are clearly valid for all periods in which there are equilibrium states—i.e., states of a homogeneous distribution of point parameters over a finite number of parts of the system. Through a precise investigation of the lessons from the "*second law*" in the sense of Ch. IV, we have arrived to the view, *that it [the second law] is not at all threatened by the prospect of an eventual reversion of real processes*, as was intended by the founders of thermodynamics and as is still the practicable [gangbare Auffassung] view. *All equations that are obtained with the help of the second law and that establish the functional dependence of different quantities on the equilibrium parameters also remain valid.*

Finally, the hypotheses of the first group also remain untouched, since they—even if they cover real processes—are independent of the direction of phenomena.

§ 43. The increase of entropy

The totality of the hypotheses of the third group implies that the entropy increases for any real process in an adiabatically isolated system.

First, let us examine the case in which the process consists in the adjustment of temperatures between two parts of a system.

Let T_α^1 and T_β^1 be the temperatures of the two parts [of the system] at the beginning of the process, and T_α^2 and T_β^2 the temperatures at the end of the process, respectively.

$$\left.\begin{array}{l}\text{Let } T_\alpha^1 > T_\beta^1 \text{ and } T_\alpha^2 = T_\beta^2 \\ \qquad\qquad \text{or } T_\alpha^2 > T_\beta^2\end{array}\right\} \tag{X}$$

depending on whether the adjustment process runs its course or is interrupted.

[65] Let the quantities of heat that are obtained during this process by each part be ΔQ_α and ΔQ_β, respectively, for which, during every elementary step of the process, it is the case that

$$\Delta Q_\alpha + \Delta Q_\beta = 0; \Delta Q_\alpha < 0.$$

In an equivalent quasi-process, both parts of the system are adiabatically isolated from one another and are thermally coupled with the appropriate external systems, through which they obtain the external quantities of heat ΔQ_α and ΔQ_β, respectively. Since the total energy of the system at the end of the real process is the same one as at its beginning, this also has to be the case for the equivalent quasi-process [Ä. Qu.]

$$Q_\alpha + Q_\beta = 0, \tag{**}$$

where

$$Q_\alpha = \int_1^2 T_\alpha dS_\alpha;\ Q_\beta = \int_1^2 T_\beta dS_\beta.$$

Among all the possible quasi-processes [Ä. Qu.] we choose the one that is most convenient for our calculations, non licet [not permitting] the one for which the temperature of each of the parts of the systems changes monotonously. In this case, we have

$$Q_\alpha = \theta_\alpha\left(S_\alpha^2 - S_\alpha^1\right);\ Q_\beta = \theta_\beta\left(S_\beta^2 - S_\beta^1\right);$$

where θ_α and θ_β represent the average values of T_α^1, T_α^2 and T_β^1, T_β^2, respectively. From the inequalities (*), it follows that $\theta_\alpha > \theta_\beta$, and therefore, from eq. (**), it follows that $\left|S_\alpha^2 - S_\alpha^1\right| < \left|S_\beta^2 - S_\beta^1\right|$ .... (***).

However, since $Q_\alpha < 0$; $Q_\beta > 0$ and both $\theta_\alpha, \theta_\beta$ are positive, we have

$$S_\alpha^2 - S_\alpha^1 < 0;\ S_\beta^2 - S_\beta^1 > 0,$$

which together with equation (***) implies that, for a real process that only consists of a heat exchange between two parts of a system, the sum of the changes in entropy of both parts [of a system] must always be positive.

2) We now examine a process during which the heat exchange between the different parts of the system is prevented, but other forms of energy can be transformed into one another or can be exchanged between the parts [of the system]. If the system is isolated from its environment, then no external addition of heat can change its entropy. Hypothesis III, 3b, however, says that such a process can be replaced by a quasi-process [Ä. Qu.], during which the [66] the system must receive a particular positive quantity of heat Q. Then, we can write

$$Q > 0;\ Q = \sum_{r=1}^{R} \int_1^2 T_r dS_r = \sum_{r=1}^{R} \theta_r\left(S_r^2 - S_r^1\right).$$

(θ_r—the average value of all temperatures achieved by this quasi-process [Ä. Qu.]). Since θ_r is always positive, and since $Q_r = \theta_r\left(S_r^2 - S_r^1\right) > 0$, in this case, we also have

$$S^2 - S^1 = \sum_{r=1}^{R}\left(S_r^2 - S_r^1\right) > 0.$$

Since the difference in entropy $S^2 - S^1$ is uniquely determined by the start and end state of the real process (since these states are equilibrium states), (Hyp. II, 2), the sign of Q will have to be the same for all other quasi-processes [Ä. Qu.]—even if the absolute values for different quasi-processes [[Ä. Qu.]] can be different.

3) Any real process, during which different energy transformations and energy exchanges occur simultaneously, can be mentally replaced by a sequence of the simple processes that we have just described, because only the final value of entropy is relevant, and this [value] is uniquely determined by the end state of the process. From above, it then follows that the entropy increases for such a process in an isolated system.

4) What happens then, if the system is isolated adiabatically but not mechanically from its environment? Will it not be able to decrease its entropy by releasing energy in the form of kinetic energy? In §22, Ch. III, we discussed a characteristic example of the transformation of pressure energy of gas into kinetic energy (case (c)). [In that discussion] Thereby, we assumed, that a part of the pressure energy in the first part of the system would remain in the form of kinetic energy in its different layers, in order to then transform into heat energy—which would increase the entropy of this part. It appears that in all cases, in which one part of a system performs work on another part against a lesser force, the quantity of heat that is thereby produced is distributed over both parts, and due to this, [67], the decrease in entropy through the performance of work is overcompensated.

N.B. It is not without interest that there are spontaneous processes, during which the initial stage [of the process] is accompanied by a decrease in entropy of the total system: e.g., a quick evaporation of liquid (e.g., like the evaporation of the ethyl ether in a vacuum) is followed by a strong cooling of the rest of the liquid, while the evaporated molecules have a larger ratio of faster velocities than would be the case in the previous state of equilibrium. Effectively, a separation of the system into two parts of different temperatures occurs, which in itself implies a decrease in the total entropy. In addition, the volume of the system increases, and this, in turn, leads to an increase in entropy, which can easily overcompensate the original decrease in entropy.

N.B. The increase in entropy is usually postulated under the general name of "second law" of thermodynamics together with an establishment of such an entropy-function whose differential is equal to the expression $\frac{\Delta Q}{T}$. This amalgamation [Verschmelzung] has to be attributed to the history of the discovery of entropy. In the previous paragraphs, it was intended to reveal the logical gulf that divides these two understandings. The form of the entropy-function depends exclusively on the holonomy of the equation $\Delta Q = 0$, this equation merely connects functions of the equilibrium parameters and their differentials, and is completely independent of the particulars of real processes. It was shown here that those hypotheses to which one can trace back the establishment of entropy and those ones to which one can trace back the properties of real processes form two distinct groups. Here, it was therefore systematically avoided to connect the term "second law" with the "increase in entropy."

[68]

§ 44. Adiabatic real processes and their mathematical description.

We now ask the following question: How can one integrate these two facts—the fact that the entropy remains constant during an adiabatic quasi-process, and the increase in entropy during a real process—using the uniform mathematical framework of the Pfaffic equations?

The example in Ch. IV, § 32 gives us a hint: a system that is not thermally homogenous can indeed change its entropy during a process for which we have $\Delta Q = 0$ constantly because, for such a system, the trajectory for which $\Delta Q = 0$ is not isentropic.

However, the process that we discussed there is only a quasi-process, which, moreover, is not "adiabatic" in the traditional sense since each of the two parts of the system is separately exchanging heat with the environment. However, the end result of each elementary step, for which we have $\Delta Q = \Delta Q_1 + \Delta Q_2$, is the same as if both parts were isolated from the environment but were thermally coupled with one another for a short time.

For a real process, however, the possibility of such an exchange of heat between the smallest neighboring parts of the system exists permanently, since this is indeed the one characteristic of a real process, i.e., that the system is not homogeneous with respect to the different parameters.

Let us restrict ourselves to the case in which the inhomogeneity is such that for each sufficiently small part of the system each point parameter is virtually homogeneously distributed. Then, as was described in Ch. § 1 5 [sic], the process can be visualized as a trajectory in the high-dimensional space R_N. In this space, however, there are no constant entropy hypersurfaces, therefore, in this space, one can connect with two points, which belong to two different diagonal spaces R_n^1, $(S = C_1)$ and R_n^2, $(S = C_2)$ of constant entropy, through a purely adiabatic trajectory. In fact: for a path element in R_N, ΔQ is given by a sum over all small subparts which, in general, have different temperatures:

$$\Delta Q = \sum_{r=1}^{R} \Delta Q_r.$$

[69]

For this reason alone, $\Delta Q = 0$ is not an isotropic equation, ΔQ is not equal to a "TdS" term. Additionally, however, each term ΔQ_r is more complicated than the expression $T_r dS_r$, if besides the terms $X_{ir} dx_{ir}$, which also occur in this case, other terms need to be added, which express the transfer of kinetic energy from a small subpart to an adjacent one and the conversion of different forms of energy into heat within each small subparts. In this case, the relation $\Delta Q_r = T_r dS_r$ will not be valid for even a small subpart.

Let us clarify the above with the most simple example of the balancing of temperatures in a solid body imaginable. Imagine two metal rods, which have temperatures T_1^0 and T_2^0 and have been laid next to one another in such a way, that they make up a single rod, which is then isolated from its environment by an adiabatic container.

The ensuing process is then governed by the equation

$$\Delta Q = \sum_{r=1}^{R} \Delta Q_r = 0,$$

where $\Delta Q_r = T_r dS_r$, and R is an arbitrarily large number of layers, into which we virtually divide the rod by imagining sufficiently close cross sections. In this case, we can assume that, relative to one another, these layers do not have any velocities. Clearly, our adiabatic equation is not isotopic, and this is consistent with the already known fact that the entropy of this system will be larger at the end of this process than at the beginning.

It is true that the current approach is valid only for particular and a highly idealized case of processes for which the assumptions of Ch. I, § 5 suffice. In this case, however, [the approach] allows one to describe every elementary step of the real process—which is not a completed process—through purely thermodynamic quantities, independent of any theory about the internal structure of matter. (The kinetic energy of the small subparts, which we thereby have to take into account, refer to the relative movement of these parts as whole bodies against one another and are not [70] the kinetic energy of the individual molecules, which is significant in the kinetic theory).

From this approach, the following remarkable result is obtained: since the equation $\Delta Q = 0$ is linearly homogeneous with respect to all differentials of the parameters of state and of the eventual velocities, it remains true even under the simultaneous change of the sign of all differentials. In other words, for this category of real processes, we have been able to mathematically analyze the possibility of a change in entropy for this category, but this analysis does not give any insight with respect to the direction of real processes! As mentioned (Ch. VI, §41), statistical theories do not provide an explanation of irreversibility either.

What has just been discussed include what seems to us to be the actual foundation of thermodynamics. In the following, we shall investigate some applications of this (as an aside, also in extremely systematic form), which are of general significance, especially in the theory of cyclic processes, which have played such a significant role in the discovery of the second law itself.

[83]

Chapter 8

Clausius Principle and Irreversibility

§ 53. The efficiency coefficient in the realization of cyclic processes

In Chapter VII, § 51, we have listed the four different statements that express the Clausius principle applied to cyclic processes. In the popular textbooks, only the first, and occasionally the second, of these are found. The reason for this is that quasi-processes are not treated as different from real processes; however, for the latter, the third and fourth statements do not hold (as long as the hypotheses of the third group are valid). We now want to examine in more detail why the directionality and the energy transformations that occur during real processes affect the validity of some of these statements.

In order for a system to carry out a real process that can be considered an approximation to a quasi-process, it is necessary that the values of the parameters of the external systems to which [the system] is coupled for this purpose are slightly different from the corresponding parameter values of the system itself. Moreover, during the process, the system cannot be homogenous with respect to the coupling parameters.

Due to these deviations from a quasi-process, the real gain in work for the environment will not be

$$A = dUp|_1^2 + dUp|_2^1,$$

where $dUp|_1^2$ represents the quasi-potential change in energy of the system for that part of the cycle, which exerts work outward, and $dUp|_2^1$ represents the same for that part, where [the system] receives work from the outside, instead, it will be a smaller value Ã. However, we should view the relationship of this quantity to the quantity of heat Q_1 that is emitted to the environment as an efficiency coefficient: [84]

$$\widetilde{\ddot{O}k} = \frac{\tilde{A}}{Q_1}$$

instead of $\frac{A}{Q_1}$. It will, therefore, be smaller than the one calculated for a cyclic quasi-process.

§ 54. Realization of Carnot processes

It will suffice for the clarification of our claims that we restrict ourselves to Carnot processes. However, how should one understand an "isotherm," if, as we have seen, at any moment of the process, the system is not in a complete state of equilibrium? Clearly, we should now interpret the two temperatures T^1 and T^2 of the two "isotherms" to be the temperatures of the two external reservoirs, assuming that they are sufficiently large, so that one can assume that their temperatures are not influenced by the coupling to the given system. The real process of our system will, therefore, run within tighter limits than the ones that we will measure for the external systems. The pressure that the system exerts toward the exterior—for that part of the cycle where it performs positive work—will, as a consequence, be smaller than in the corresponding ideal Carnot process; the pressure against which it will act, in order for the corresponding geometrical parameter to change (in the easy example, so that the piston can move), will have to be even smaller, with the result, that the positive work that has been exerted will be less than in the ideal case. One can easily convince oneself that for the part of the cycle during the system will do negative work, these relationships will be reversed. Therefore, the total positive work done will be less than calculated for the ideal Carnot process. If the temperature- or pressure differences between the given system and its environment are not very small, then—in principle—one cannot guarantee that the system, once it has received a positive quantity of heat Q_1 from the environment, has not given the same value $Q_2 = Q_1$ to the environment and that therefore the exerted work is zero. We see, therefore, why the third form of the Clausius principle becomes invalid!

We leave it to the reader to convince themselves that the validity of the fourth form [of the statement] can also not be guaranteed anymore.

[85] Let us turn our attention to the consequences that the inversion of all phenomena would have and, as an example, let us observe the inversion of a real Carnot process, in which heat is converted into work. For the inverted process, two things would occur: the transfer of a particular amount of work A into heat, and the transfer of a quantity of heat Q_2 different from zero from a lower temperature T^2 to a higher temperature T^1. If for the case of the directed process, none of the three quantities can be zero (first form of the Clausius principle), then they can also not be zero for the inverse process. This, however, amounts to the fourth form of the Clausius principle.

One can easily convince oneself that the third form can be recovered from the examination of the inversion of a Carnot process during which heat from a lower temperature is transferred to a higher temperature, to which the second form of the Clausius principle applies.

In contrast, during the inverted trajectories of all phenomena, the first and second forms of the Clausius principle prove to be uncertain, since the conversion of heat into work in this case is accomplished by means of a Carnot process that is the inversion of a process during which work is done on the system by the environment and during which it is not certain whether a non-zero quantity of heat Q_2 is brought from a reservoir of lower temperature to a reservoir of higher temperature. The analog is true for the case in which heat is brought from a lower temperature to a higher one: it will be uncertainty, whether during this the system will receive heat from its environment. Therefore, the two first forms of the Clausius principles appear to be invalid.

Concerning the efficiency coefficient, for the case of inverse trajectories, the value $\frac{T^1-T^2}{T^1}$ will not represent the upper but the lower limit. It will be the case that

$$1 > \widetilde{\ddot{O}k} > \frac{T^1-T^2}{T^1}$$

[86]

§ 55. The real role of irreversibility in the theory of efficiency coefficients.

The speculations about inverse processes—regardless of whether one believes them to be possible or not—are in any case instructive in so far as they show us the real role that the irreversibility of real processes plays in the Clausius principle: without the one-sidedness of all real processes for a given period (i.e., if any real process could run once in one direction and another time in the opposite direction), then absolutely nothing would remain from this principle (if applied to real processes)! In time periods during which one-sidedness is present, two of the four statements hold up; which ones these are—the first two or the last two—depends on the direction of real processes. On the existence of the [integrating denominator] I.N. of ΔQ, however, irreversibility does not have any influence.

[125]

Appendix III

Boltzmann's H-Theorem

§ 1. The equating of the H-function with the S-entropy-function, which was discussed in appendix II, is based on the assumption of the Boltzmann–Maxwell distribution ("B-M")

$$f_Z = f_0 e^{\frac{-\varepsilon_Z}{kT}} \qquad (*)$$

for a system at equilibrium. Boltzmann himself derived this formula from the equation:

$$\frac{dH}{dt} = -\alpha \int (f_1 f_2 - f_3 f_4)(\log f_1 f_2 - \log f_3 f_4)\, d\Omega \qquad (\mathrm{H}_1)$$

which describes the temporal change of the H-F [H-function] of an isolated system, if this system is not in an equilibrium state and if it changes its state in response to the movement of its constituting molecules. Since both factors in the integral in the formula above change their sign simultaneously, from eq. (H_1), it follows that an isolated system can perform only such spontaneous processes for which its H-F strictly decreases, unless it has reached a distribution that satisfies the equation

$$f_1 f_2 - f_3 f_4 = 0 \qquad (\mathrm{H}_2)$$

Both the B-M distribution, and, in turn, eq. (H_2) follow from this equation, the latter implies that, after the attainment of the B-M distribution, the thermodynamic state of the system remains unchanging, that it has reached its equilibrium state. This is the content of the "H-Theorem" in its original form. It, therefore, delivers two important results: I) the distribution of molecules over the coordinate- and over the momentum-states in the equilibrium state of the system and 2) the claim of the one-sidedness, i.e., the "irreversibility," of processes in an isolated system. If one extends the equating of H-F with the entropy S for all possible (including turbulent) processes, then one obtains an explanation for the increase in entropy [126] during all real processes, as follows from the hypotheses of the III group (Chapter 6 of this book), if one only accepts the first form of these hypotheses, as is indeed customary in physics. Then, these hypotheses appear to be proven from within the framework of kinetic theory.

§ 2. Shortly after its publication, objections were raised against the H-Theorem, most importantly, Zermelo noted that the premises from which Boltzmann's calculation started (i.e., that the totality of the molecules of the system had constant energy and that these molecules were subjected to conservative forces only) have as a necessary consequence that, in time, the system will return arbitrarily closely to each state that it had previously taken up, in other words, that the trajectories of the processes in such a system should not be one-directional, but quasi-periodic (if not even occasionally exactly periodic, up to a relatively negligible number of cases in which they can be asymptotic). Indeed, an assumption has been made during the derivation of eq. (H_1), whose general validity could not be proven: the "Stosszahlansatz" ("STA"):

$$\Delta f_z = \Delta t \beta f_z \qquad (H_3)$$

which states that the number of molecules of a particular kind Δf_z (defined by their coordinates and their moments) that, within a given small time interval Δt, change their velocities through collisions with other molecules, is proportional to the number of molecules of this kind f_z that were present at the beginning of this time interval.

If any changes of the H-F were to be quasi-periodic, then it would clearly be impossible to satisfy this assumption [STA] without exceptions.

Boltzmann also explained that this assumption would provide only the most likely equation for a given number of collisions; that, certainly, sometimes there could be deviations from it; that, however, compared to the cases in which the assumption was satisfied, these were extraordinarily rare, such that the equation (H_1) would be overwhelmingly probable. (This was the signal for the now widespread use of the "probability"-approach and for the pushing back of the deterministic-mechanical premises upon which [127] the realization of the quasi-periodic character of all phenomena was founded!).

Under the new meaning of the H-Theorem, the temporal course of the H-F of an isolated system was represented by a curve that is composed of extraordinarily long sections of straight lines, which lie on the minimal height H_{mix}, and which are separated from one another by, more or less, non-homogeneously distributed peaks.

The possibility of an equilibrium [state] that would last indefinitely should then be given up: similarly, the B-M distribution could not persist indefinitely; however, it could be seen practically as a rest state—as an "equilibrium state," since it could be reproduced itself so extraordinarily often.

§ 3. If the H-F was still identified with the thermodynamic entropy S, then one would have to revise the foundations of thermodynamics, and, in particular, one would have to allow that (in the language of this book), in addition to the first form of the hypotheses of the III group, for certain time intervals, the second form should be also valid.

This was already something unusual; however, another difficulty was added to this. It should be said immediately, with emphasis, that this difficulty is based on a misunderstanding: since the discovery of the function "entropy"-function by Clausius, one was used to tie the existence of the integrating denominator ("IN") of the expression

$$\Delta Q = dU + \Delta A \qquad (H_4)$$

which was discussed in Ch. (IV), to the irreversibility of real processes; the possibility of the existence of reversed trajectories of real processes was therefore seen as a threat to the "second law" of thermodynamics, whereby, under this name, one understood without differentiation, both the creation of entropy (given as a function that was defined by the equation

$$\frac{\Delta Q}{T} = dS \qquad (H_5)$$

) as well as the increase in entropy in the case of real processes.

[128] In Ch. VI of this book, we have seen that equation (H_4) is valid only for quasi-processes (for real processes, we should replace it with the equation

$$\Delta Q = \Delta U + \overline{\Delta A} \qquad (\overline{H_4})$$

). The IN only refers to equation (H_4), its existence is independent of which direction real processes run. Therefore, the possibility of the reversion of real processes, which is implied by Zermelo's objection, is not a threat to those equations of thermodynamics that are based on the existence of the IN.

Nevertheless, Clausius' suggestion was so influential that even today, the "second law" is identified with the law of the "increase in entropy" by physicists. Even Carathéodory, who contributed so much to the clarification of thermodynamic concepts, formulated his "Axiom II," to which he reduced equation (H_5), in such a way that it encompassed both quasi-processes and real processes. (When I made him aware of the fact that it would be less ambiguous, if his axiom were only applied to quasi-processes, a longer correspondence started between us; even when, in principle, he had conceded to my point of view, he did not want to change the formulation of the axiom, since it seemed to him to be more "physical").

§ 4. Due to the prejudice mentioned in § 3, a succession of efforts (building on the new meaning of the H-Theorem, therefore building on the new form of the H-curve) to look for such interpretations, under which the decrease in H would seem more probable than its increase, began. In the best case, these could only be explanations of the subjective attitude of the researcher, who, if thinking about a state of the system that is different from a state of rest, finds it difficult to imagine that the system would subsequently depart even further from this state of rest. (Among these kinds of work one can also count the article by P. and T. Ehrenfest; "*Üeber [sic] zwei bekannte Einwände genen das Bolltzmannsche [sic] H-Theorem, Phys. Zschr.* 8, [1907], p. 316).”

Under no circumstances could one deliver a proof of something that, from the beginning, was ruled out by the accepted form of the H-curve. (See T. Ehrenfest-Afanassjewa, *On a misconception in the* [129] *probability theory of irreversible processes, Acad. Amsterdam* Vol. XXVIII [1925] N 8–9).

The thought of Boltzmann, that the STA was overwhelmingly probably, seemed very seductive. In fact, one can assume this without contradicting any of the aforementioned objections to the original form of the H-Theorem! There is no reason to deny that the H-theorem is overwhelmingly probable. But this need not be confused with the claim that a reduction of the H-F is more probable than an increase: the overwhelming majority of time intervals during which STA is realized coincides with those intervals during which the B-M distribution holds (the concept of the “evenly disorderly” distribution, on which Boltzmann based his STA, seems to correspond best to the B-M distribution). Among all the cases during which the H-F changes its value, it is clearly that at least half [of the cases], namely, those time intervals during which H increases, are inconsistent with the STA. Therefore, we are justified to use the calculations of Appendix II in order to explain equation (H_5), which expresses the actual content of the “second law,” thereby starting from kinetic theory, and, as a result, we can deny—according to the same theory—the permanent irreversibility of real processes.

§ 5. It was inevitable that some physicists were unhappy with the bogus proof [Scheinbeweis] of the tendency to decrease the H-F, as discussed in § 4. So, the objection to the assumption of a mechanical determinism for all processes by Zermelo was well founded. One expected that the interpretation of the phenomena in terms of quantum mechanics, where determinism is fundamentally given up, would lead to a reconciliation with irreversibility. All that could be achieved, however, was the claim that the function that in this interpretation was supposed to be analog to the H-F could only … increase extraordinarily seldomly. This, however, is not fundamentally different from what the kinetic theory had produced. One can well say that this was also to be expected a priori: if one assumes that the values of a quantity are determined by chance and that the frequencies of the different values are proportional to the relative probabilities of these values, then it follows from this that during an infinite time interval, each value for which the probability is different from zero [130] will have to recur, that, therefore, any changes in this quantities will have to be quasi-periodic.

§ 6. If one accepts the conclusions of the last paragraphs, then there arises one question, which has a great meaning for the progression of physics: all of our experiments and all predictions of the trajectories of phenomena are based on the assumption of irreversibility (we expect that the temperatures of two bodies that touch one another will become one and the same, we do not expect that, within a foreseeable time interval, they will assume different temperatures again, and vice versa); how is this to be reconciled with the quasi-periodic trajectories [of phenomena]? One can answer this question within the framework of kinetic theory. According to this theory, every state of a system is completely determined by its previous state; and this is also the case for an arbitrarily large part of the observationally accessible world; (If one suddenly moves the middle piston of example 2 Ch. 1 §1 to one side, so that behind it one creates empty space, then the expansion of the gas into the enlarged space has to follow for purely deterministic reasons, and it is unconceivable that, after this, the gas could move all of its mass immediately in the opposite direction). Now, we know from everyday experience that the world that surrounds us is currently in such a state; that, as all processes proceed in the direction that is familiar to us, it is inconceivable that an arrangement of molecules that would cause reversed trajectories of the phenomena would suddenly occur. One can formulate it thusly: the macroscopic laws have a certain "inertia"; they cannot change unexpectedly.

If one asks: what should one think about the fact that we were born into such a period during which changes in entropy during real processes always lead to an increase in entropy and neither to a decrease nor to an occasionally change in direction? Then this is the answer: it is not a coincidence, our organisms are constituted such that they can only evolve and develop under the circumstances that exist during the current state of the world. What it will look like, when the period of reversed trajectories[of phenomena] begins, [whether] combinations of molecules that are similar to our bodies will then occur, while experiencing changes [131] that run in the opposite direction—this one cannot answer easily: one should not forget that in all our discussions there has been no mention of exactly reversed trajectories of phenomena: "quasi-periodic return" only means that, after a finite amount of time, the system has to come close to a state that it had taken up earlier; the order in which states follow upon another is not determined by this statement.

If, however, we consider a chance to be the ultima ratio of all phenomena, then we do not have any answer to the question we have asked above: in principle, there is no reason to exclude arbitrary jumps in state! Even worse, it is astonishing that the world crawls so unexpectedly slowly toward the overwhelmingly probable state of rest (of general homogeneity), if, at any moment, it could jump to it!

Translation from an article appearing in *Zeitschrift für Physik*, **33**, 933–945, (1925)

Note: in this translation, the symbol *A* (Arbeit) has been replaced by *W* (work), an apparent misprint in the last equation (dQ = TAS) of section 2 has been corrected, and the name Caratheodory to Carathéodory.

On the Axiomatization of the Second Law of Thermodynamics

Tatiana Ehrenfest-Afanassjewa

Abstract:

A conceptual analysis of the second law of thermodynamics is carried out in connection with an investigation of Carathéodory. It will be shown that the law can be reduced to two groups of logically independent axioms. One of these groups applies to the properties of bodies in equilibrium and the other concerns the laws of irreversibility. This shows that the Thomson principle and the Clausius principle do not have the exact same axiomatic structure.

Carathéodory's[8] attempt to axiomatize thermodynamics seems to be a unique, yet particularly valuable attempt. This attempt can be assumed to be generally

[8]C. Carathéodory, Untersuchungen über die Grundlagen der Thermodynamik. Math. Ann. **67**, 335, 1909.

known after the explanation that M. Born[9] has given of it in the "*Physikalische Zeitschrift.*" Regarding what we are concerned with here, Carathéodory has achieved the following: he has reduced the existence of the integrating denominator for the expression $dQ = dU + dA$, and therefore also the existence of Entropy, to a simpler axiom than the second law ("Axiom II"). By invoking the theory of Paffian equations (which he also complemented with an essential Theorem) he has provided clarity to the existing relation that was looked for in vain in the previous derivations of the integrating denominator dQ. He has further shown according to what general mathematical reasons the integrating denominator has to have a part that can be separated as a factor, which is a pure function of temperature and represents absolute temperature.

These results have served as a starting point for this study. However, the approach here is different from Carathéodory's, most of all concerning the role of irreversible processes. The aim, here, is to obtain a deeper understanding of the second law by concentrating exclusively on reversible processes.

§ **1**. We first observe that a thermodynamic system in equilibrium is defined in terms of a finite number n of parameters, that we will call $x_1, x_2, \ldots x_n$.

Following Carathéodory, we will call a continuous series of equilibrium states a "quasi-static" change of state.

We assume that the concepts of "internal energy," "the quantity of heat added to the system," "temperature" are known.[10]

We have

$$dQ = dU + dW = Y_1 dx_1 + Y_2 dx_2 + \ldots + Y_n dx_n,$$

where dQ is the quantity of heat added to the system in an infinitesimal quasi-static change of state, dU is the change in internal energy, dW is the work exerted by the system and Y_i $(i = 1, 2, \ldots, n)$ are functions of the parameters of state.

We say a system is "thermally homogeneous" if all of its parts are coupled to one another, in such a way that at every moment of the change of state they have one and the same temperature.

We say a change of state is "adiabatic," if the infinitesimal changes in parameters satisfy, without exception, the equation

$$Y_1 dx_1 + Y_2 dx_2 + \ldots + Y_n dx_n = 0$$

throughout the process.[11] Otherwise we call the change of state "not-adiabatic."

---

[9]M. Born, Kritische Betrachtungen zur traditionellen Darstellung der Thermodynamik. Phys. ZS. **22**, 218 und 282, 1921.

[10]The critically examined determination of these concepts can be found in the above-cited work of Carathéodory, and this discussion will be ignored in this paper. See also C. Carathéodory: Über die Bestimmung der Energie usw. Berl. Ber. 1925.

[11]This does not imply that the system has to be adiabatically isolated from its environment. Think of a system composed of two gases that are separated from one another by a moving and adiabatically isolating piston. One can let this system undergo a change of state, throughout which $dQ = dQ_1 + dQ_2 = 0$ is satisfied constantly, but $dQ_1 = -dQ_2 \neq 0$, and both parts of the system, that are

§ **2**. **Axiom A** (Entropy Axiom). If the integral $\int_1^2 dQ$ is non-zero along a quasi-static path that connects states 1 and 2 of a thermally homogeneous[12] system, then the system cannot be brought from one of these states into the another along an adiabatic quasi-static path.

It follows from this axiom that, in an arbitrarily small neighborhood of every state, there are always adiabatically inaccessible states.[13] As Carathéodory has shown, the existence of the integrating denominator of dQ and of the entropy can be derived from this fact.

With this result, however, one has only accomplished that the expression for dQ can be reduced to the same form as the one for the other terms of the identity.

$$dQ - Y_1 dx_1 - Y_2 dx_2 - \ldots - Y_n dx_n = 0,$$

namely to the form of a product of the functions of state with a differential:

$$dQ = TdS.$$

We can say that the Axiom A suffices to derive the theorem of the existence of entropy. The second law of thermodynamics is however not yet exhausted.

What is the second law? We know two typical formulations thereof: one by W Thomson "no perpetual motion machine of the second kind is possible," and one by Clausius: "in a cyclic process, it is impossible to transform heat into work without simultaneously extracting a quantity of heat from a warmer to a colder body."

We will see, that these two formulations are not equivalent.

§ **3**. In order to derive the law of W. Thomson, it is necessary to add two more axioms to Axiom A.[14]

Axiom B (Coupling axiom). Only one kind of thermal coupling is possible.

We mean by this that when two systems are coupled with one another in such a way that—maintaining equilibrium—they can exchange heat, the exchange of heat is only possible if the two systems have the same temperature[15] (and not, instead, any other function of state), and

adiabatically isolated from one another, respectively, take this quantity of heat from and transfer this quantity of heat to their environment.

[12]See §5 of this paper.

[13]One means "unattainable by means of a quasi-static adiabatic process," as only this is necessary for the definition of entropy.

[14](Annotation made in the correction of this paper): it would have been better to let the axiom of uniqueness precede the coupling axiom.

[15]A counterexample is given by the case of the coupling "after the term dW," e.g., by the coupling of the work done in order to change the volume of a gas (Druckarbeit): two systems can exchange work in such a way that either the pressures are equal, or otherwise only all the forces that they exert on one another are equal without the pressures being equal.

Due to this state of affairs, one can extract work from a reservoir of constant pressure (and consequently transform it into heat):

Axiom C (Uniqueness axiom). The integral $\int dS$ along a closed path is always equal to zero.

Axiom B is necessary, otherwise one could obtain a value $\int TdS \neq 0$ along a closed path even if the system exchanges heat with only one heat reservoir of constant temperature T_0. This means that, in a thermodynamic cycle, one could extract heat from a reservoir of constant temperature and transform it into work.[16]

Axiom C is also necessary, because without this axiom one could conceive of a thermodynamic cycle in which the system exchanged heat while having the same constant temperature T_0 as the heat reservoir the value of the integral $\int dQ = T_0 \int dS$ would be non-zero, because of the possibility that $\int dS \neq 0$.[17]

Axioms A, B, and C are also sufficient for the requirement that we need more than one heat reservoir to obtain work from heat in a thermodynamic cycle: in fact, using these axioms, one finds that $\int dQ$ becomes zero when one allows heat exchange with only one reservoir.[18]

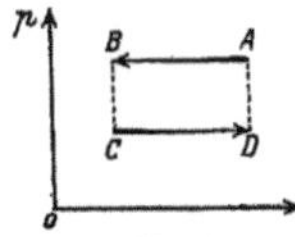

Fig. 1.

Let the system has to be a quantity of gas isolated by two movable pistons *a* and *b*, which are of the same size. The reservoir consists of an infinitely large quantity of gas that is enclosed by two pistons 1 and 2 of different size. One can, alternatively, let piston 1 press on piston *a* and piston 2 press on piston *b*. In the first case, we have $dW = p_1 Sdl$ for the System and $dW' = pS_1 dl'$ for the reservoir, where S and S_1 are the cross sections of piston *a* and 1, respectively, p_1 and p are the pressure values of the system and of the reservoir and dl and dl' the displacement of both pistons. In the second case, we similarly have $dW = p_2 Sdl$ and $dW' = pS_2 dl'$.

Clearly, we will have $dl = dl'$. Since, however, we need to have $dW = -dW'$, then $p_1 S = pS_1$ and $p_2 S = pS_2$.

With the help of this couplings, we can let the system carry out the cycle represented in Fig. 1. The quantity of work obtained in this cycle is

$$-\left\{\int_A^B p_1 Sdl + \int_C^D p_2 Sdl\right\} = -\left\{p_1 \int_A^B dv + p_2 \int_C^D dv\right\} = -(p_1 - p_2)(V_B - V_A).$$

This quantity of work is positive and is obtained from a single reservoir of constant pressure. P. Ehrenfest has brought the significance of the ambiguity of the types of coupling in this context to my attention.

[16] See the analogy in the previous footnote.

[17] It is easily accepted that for $T \neq 0$ the coefficients Y_i do not have any singularities, which could be a reason for the ambiguity of the function S. If the value of the coefficients Y_i vanish when the absolute temperature has the value zero, then the differential dS can produce an indeterminacy of the type $(\frac{Y_i}{T} = \frac{0}{0})$, for which S would become ambiguous. Moreover, if the absolute temperature has value zero, this would also give a reason for the breakdown of the formulation of Thomson's formulation of the law despite the validity of Axiom C. In fact, if one could set to zero the temperature of the colder reservoir in a Carnot cycle, then one would have $Q_2 = \int dQ = 0 \int dS = 0(S_B - S_A)$, where the difference of the entropies would be finite.

[18] Assume we reach the value zero for absolute temperature. Two types of attitude can be held toward this value for absolute temperature: either one declares this value is unattainable by means of a special axiom, and in this case Thomson's (and Clausius') principle holds for "for all attainable

§ **4**. Axioms A, B, and C are necessary in order to obtain Clausius' Principle,[19] however, they are not sufficient. In order to have a complete justification of this principle, the following axiom is also necessary:

Axiom D (Temperature Axiom). The absolute temperature has one and the same sign for all states.

As a matter of fact, without this condition, one could carry out a Carnot Cycle which would contain one isotherm with positive and the other with negative absolute temperature. In this case, it would be possible to extract heat from both reservoirs[20] and in the process of transforming heat into work, no heat would go from a higher to a lower temperature, as is required by Clausius' principle.

It is taken as self-evident that the absolute temperature has an invariable sign. We can see from the above discussion, how this property is essential for Clausius' principle.[21]

On the other hand, Axiom D is not necessary for Thomson's principle, as this principle only says that in order to turn heat into work one needs more than one heat reservoir.

---

states," or one declares that the principle holds "for all processes in which the value zero for absolute temperature is not attained."

[19]The fact that the existence and the meaning of entropy contained within this principle follows from its derivation from the second law, which Clausius provided himself. Further, it follows from axiom A and C that two adiabatic paths W_a and W_b that follow a part of an isothermal path from the two ends a and b, and cannot have any point in common. If one wants to perform a cycle, that contains the isotherm ab as one of its parts, then the two adiabatic paths W_a and W_b have to be connected by a non-adiabatic path. And although this non-adiabatic path might be an isotherm (cd), then it cannot have the same temperature as ab (otherwise we would again have

$$\int dQ = T_1 \int_a^b dS + T_1 \int_c^d dS = T_1(S_b - S_a) + T_1(S_d - S_c) = 0$$

because $S_b = S_c$ and $S_a - S_d$). In order for the heat to be emitted to a reservoir that has a different temperature $T_2 \neq T_1$ along the second path—as is required by Clausius' Principle—it is impossible for this reservoir and the system to have different temperature. Axiom B is therefore necessary also in this case.

[20]Analogously to how one can extract work from both "work-reservoirs," as is represented in Fig. 2. Imagine a coil that tends to expand with higher temperatures and to contract with lower temperature within the lengths ($l_1 l_2$).

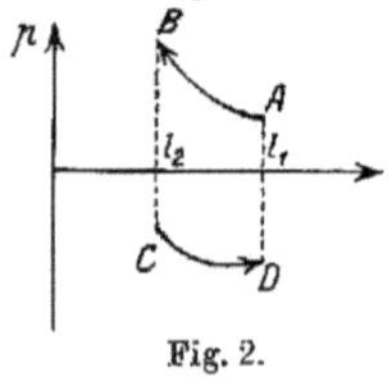

Fig. 2.

[21]In the interpretation of absolute temperature in classical statistical mechanics, it naturally follows that this quantity can only be positive, in so far as it is the mean kinetic energy of the molecules. However, whenever one is compelled—e.g., because of quantum theory—to deviate from this interpretation, Axiom D requires another special statistical interpretation.

We, therefore, see that the two principles do not say the same thing and that one cannot derive neither Thomson's nor Clausius' principle from the existence of entropy alone.[22]

§ **5**. Remark. Axiom A is valid only for thermally homogeneous systems. A system that is made up of two (or more) parts that are adiabatically separated from one another, but coupled by pressure and have different heat capacities [939], may very well go from one state to another quasistatically and adiabatically, even if these two states cannot be linked along an adiabatic path (for which $\int dQ \neq 0$). Indeed, the expression

$$dQ = c_1 dT_1 + p dv_1 + c_2 dT_2 + p dv_2$$

does not have an integrating denominator, even if both parts of the system are ideal gases.[23] If there were such states in the immediate vicinity of every state, then the above expression would have an integrating denominator—as Carathéodory has shown.[24] From this we see that by changes of state that satisfy the equation $dQ = 0$ the system can reach *arbitrary* states in the vicinity of its initial state, even those states that can otherwise be reached when $dQ \neq 0$.

Nevertheless, even such a system obeys Clausius' principle. This can be easily derived from the condition that $\int dQ \neq 0$ for a closed cycle, combined with the fact that, along the same path we must have $\int \frac{dQ_1}{dT_1} = 0$ and $\int \frac{dQ_2}{dT_2} = 0$.

§ **6**. Four equivalent formulations of the "second law for quasi-static processes" follow simultaneously from axioms A, B, C, and D.[25]

In a quasi-static cyclic process

1. heat cannot be transformed into work, without simultaneously transferring an equivalent quantity of heat from a warmer to a colder body;
2. heat cannot be transferred from a colder to a warmer body, without simultaneously transforming an equivalent quantity of work into heat;
3. work cannot be transformed into heat, without simultaneously transferring an equivalent quantity of heat from a colder to a warmer body;
4. heat cannot be transferred from a warmer to a colder body, without simultaneously transforming an equivalent quantity of heat into work.

The last two formulations show that the second law alone does not imply any dissipation of energy. One can learn even more by considering these formulations

---

[22]This also explains why the various attempts to construct analogies to the second law could not be carried out completely.

[23]The condition for integrability is, in fact

$$Y_1\left(\frac{\partial Y_2}{\partial x_3} - \frac{\partial Y_3}{\partial x_2}\right) + Y_2\left(\frac{\partial Y_3}{\partial x_1} - \frac{\partial Y_1}{\partial x_3}\right) + Y_3\left(\frac{\partial Y_1}{\partial x_2} - \frac{\partial Y_2}{\partial x_1}\right) = 0$$

for the Pfaffian expression for three parameters. In our case, however, this condition is violated when $c_1 \neq c_2$.

[24]L. c.

[25]The note by Arthur E. Ruark, "The Proof of the Corollary of Carnot's Theorem" (Phil. Mag. 49, 584, 1925) has appeared during the writing of the present work. In that work, Ruark also points at the possibility of the latter two formulations.

simultaneously, if we want to give an account of the relationship of the second law to the irreversibility of natural processes.

§ 7. Imagine, next to the "normal" world, a "reversed" one, in which all irreversible processes have the opposite direction. Carry out one and the same Carnot cycle in both worlds and imagine that in both cases the two heat reservoirs are not completely isolated from one another. (For the sake of simplicity, assume that this is the only cause of the irreversibility of the process; this will suffice to clarify the essential aspect of our example.)

In the first case, a quantity Q' of heat will seep from the warmer to the colder reservoir during the cyclic process. The actual quantity of heat that would be transferred to the colder reservoir will not be equal to the quantity Q_2 that would be transferred during the cyclic process if the reservoirs were completely isolated (with a given dW). Instead it would be $Q_2 + Q'$; likewise, Q' would be extracted from the warmer reservoir in addition to the Quantum Q_1. We obtain therefore

$$\frac{dW}{Q_1 + Q'} = \frac{(Q_1 + Q') - (Q_2 + Q')}{Q_1 + Q'} < \frac{Q_1 - Q_2}{Q_1}.$$

The quotient

$$\frac{Q_1 - Q_2}{Q_1} = \frac{T_1 - T_2}{T_1},$$

therefore appears as the *upper* bound of the economic coefficient.

In the second case, a quantity Q'' of heat will seep from the colder to the warmer reservoir during a cyclic process, and if we require the previous value dW and start from the same temperature T_1, we will obtain

$$\frac{dW}{Q_1 - Q''} = \frac{(Q_1 - Q'') - (Q_2 - Q'')}{Q_1 - Q''} > \frac{Q_1 - Q_2}{Q_1},$$

so that

$$\frac{Q_1 - Q_2}{Q_1} = \frac{T_1 - T_2}{T_1}$$

appears as the *lower* bound of the economic coefficient.

Because of the possibility to directly transfer the heat from one heat reservoir to another in an irreversible way, in addition to a reversible way (so with the transformation of one part of the quantity of heat Q_1 into work or the other way around), the second half of the formulations of Clausius' law lose their validity[26]; in the "normal" world the first two formulations are valid, in the "reversed" world, the latter two.

[26]Because Q'' can be equal to or even bigger than Q_2; as a consequence, formulations 1 and 2 would not be valid in the "reversed" world.

All the properties of the states of equilibrium and of the quasi-static changes of state are the same in both worlds, as follows from axioms A, B, C, and D.

§ **8**. However, on closer inspection, we recognize that the whole theory of quasi-static changes of state would be valid even if the sequence of states of a system during an irreversible process were actually reversible. It is useful to understand the difference between the concepts "unattainability" and "irreversibility" and their relation to the whole theory.

In so far as the expression "unattainability" is used to justify the claim that dQ has an integrating denominator, it refers to an equation (the Pfaffian equation:

$$Y_1 dx_1 + Y_2 dx_2 + \cdots + Y_n dx_n = 0),$$

that does i contain time as a variable. This equation only defines a relation between changes in the parameters that correspond to the respective adiabatic changes of state; one can read off from this equation which consecutive equilibrium states can lie on an adiabatic path. However, one cannot read off from this equation the direction in time in which this path will have to run, be this in both directions or just one. Two states that lie on two different sides of one state along an adiabatic path will be "attainable," according to this equation—even if this path is "irreversible," according to the temporal sequence. Whether a particular state can be attained adiabatically quasistatically from a given one, only depends on the coefficients of this Pfaffian equation, and nothing else. So we conclude first that for the existence of entropy it is irrelevant whether the quasi-static processes are reversible or not.

Now consider the sequences of states traversed by a system in an irreversible process—understood in the usual sense—and for the moment ignore that these sequences are in fact irreversible. We call these sequences "non-static" changes of state.

The state of a system at every instant of a non-static process is not defined by the same number n of parameters $x_1, x_2, \ldots x_n,$ as is the case for quasi-static processes: as soon as a non-static process begins, the homogeneity regarding some of these parameters is destroyed, one immediately has a temperature, pressure, or concentration drop.

[Often (for turbulent processes) one cannot ascribe a precise value for these parameters even for the smallest part of the system. For slow processes (especially when the system is made out of solid bodies or fluids) one can, however, ascribe (in good approximation) a precise value to these parameters to each point of the system. These values, however, vary for each point. These are therefore states that are defined by infinitely many parameter values.]

If we think of a system in a state of equilibrium to be represented by a point in an n-dimensional parameter space, then we have to say that as soon as a non-static process begins this point will have to leave the parameter space. [If we approximate the different parts of the systems with a continuum, we can speak, for the case of slow, non-static processes, of an "infinitely dimensional" space, from which the n-dimensional space forms a subspace and the point that represents the system leaves this n-dimensional space. This point will describe a path that will have its starting point and end point in this n-dimensional space.]

The non-static changes of state are therefore not represented in the n-dimensional space, for which the Pfaffian equation is valid.[27] As a consequence, they have no relevance for the question concerning whether a path that obeys this equation from a particular point in this space to another point is attainable or not.

We therefore secondly conclude that the potential reversibility of non-static processes would have no influence on this unattainability.

We, therefore, see that the quasi-static, adiabatic unattainability of particular states that are near a given state does not have anything to do with the question concerning the irreversibility of any process; in particular, this unattainability is not a consequence of the irreversibility of non-static processes. It is for this reason that the formulation of Carathéodory's "Axiom II" is replaced with "Axiom A" above.

The second law for *quasi-static* changes of state[28] would therefore also not be violated if one could find a way to make non-static processes reversible. However, the law would lose its meaning for processes that depart greatly from quasi-static processes: all four formulations would be violated, and the economic coefficient would not have any bound. This means: the irreversibility of non-static processes has no relevance for the existence of entropy, but it is fundamental for the validity of Clausius' principle *for real processes.*[29]

§ 9. The following remark is allowed here: Thermodynamics would gain much more conceptual clarity if one fixed by explicit axioms: 1. the conditions for the realization of non-static processes, 2. their irreversibility, and 3. the determination of their direction, and emphasized them as a third group of axioms (axioms of irreversibility) in addition to the second group: Axioms A, B, C, and D (to which one could also

[27] One could think, for the case of slow processes, of expressing their energy balance in terms of an equation with infinitely many terms, which, however, would not be holonomic.

[28] And for those changes of state that depart only slightly from quasistaticity.

[29] Let us compare the thermally non-homogeneous systems that go through quasi-static changes of state discussed in §5 with thermally non-homogeneous systems that go through non-static changes of state. Let us not be disturbed by the fact that, for the latter processes, the expression for dQ would contain infinitely many terms. What these two groups of processes have in common is that for both groups, the expression dQ is not holonomic, i.e., it does not have an integrating denominator. As a consequence, both systems could perform a cyclic process, in which for one part of the path heat is extracted from a single reservoir of constant temperature, and in which for the remaining path the total quantity of heat dQ obtained by the total system is zero. The fundamental difference, however, is that, on this kind of "adiabatic" path, a system from the first group could not be isolated adiabatically from its environment (otherwise—because of adiabatic isolation with each other—each of its components would not obtain any heat, and this would reduce excessively the variability of the parameters)—therefore, there should be other heat reservoirs in addition to the first one. Among these other reservoirs there also have to be ones (as one can show), that, on the whole, take up heat from the system. On the other hand, a system from the second group can remain adiabatically isolated from its environment during the whole part of the adiabatic section of the path of the cycle, as each of its parts can obtain the required heat from its neighboring parts. This is the reason that, with regard to non-static processes, something from Clausius' Principle will remain only if the non-static processes are prohibited in at least one direction.

add Nernst's axiom) and the first group of axioms, that serve for the definition of internal energy, quantity of heat, and temperature (Axioms of the First Law).

The "*Second Law for all real processes*" would result from a combination of the two latter groups.

The relationship of this third group of axioms to the second group can also be expressed by the following scheme:

a) All formulations of Clausius's law are valid for quasi-static processes, independently from the irreversibility of non-static processes.
b) All four formulations lose their validity for the totality of quasi-static and non-static processes if the latter are also reversible.
c) Two of the four formulations are saved for the totality of quasi-static and non-static processes if the latter are irreversible. Whether it is the first two or the last two that are saved depends on the directionality of the allowed non-static processes.
d) If the quasi-static processes are also irreversible, then this would not at all affect the existence of entropy, as it would not modify the coefficients Y_i of the Pfaffian equation considered above. Because of the same reason, it would also not yield a new integral of this equation. However, it would not make sense anymore to talk about cycles and everything that has to do with them.[30]

§ **10**. The comparison of remarks a, b, and c can also contribute to reconcile the cleft between classical thermodynamics and the kinetic theory of thermodynamic phenomena; this comparison shows how much would remain from thermodynamics if one would want to expand it to the periods in which processes run in the opposite course.[31]

Leiden, July 18 1925.

---

[30]Remark d has to be mentioned—despite its complete abstraction from physical aspects—as this remark illuminates the mathematical relations contained in the second law from a new perspective and it shows once more the very different role that the axioms of the second and third group play in thermodynamics.

[31]One believed, at the beginning, that the incessant increase of entropy (with which one identified the whole second law) had really been proven through the Boltzmannian H-Theorem on the basis of kinetic interpretations, while in classical thermodynamics the second law appears only as a postulate. On closer inspection, however, one sees that both theories need to be founded on axioms that, each in their own language, actually say the same thing: equalization of temperature, pressure, etc., in classical thermodynamics; the *Stoßzahlansatz* in kinetic theory. Both theories, therefore start more or less from the same axiom. However, in the kinetic theory—now that it has been put under the magnifying glass—it does not appear to be so unconditionally valid as it is for classical thermodynamics; indeed, what is more: the periods in which it [the axiom] is not satisfied and in which all phenomena show an opposite direction, are equally probable as those that show the "normal" course of direction. A discrepancy between classical thermodynamics and the kinetic theory appears in the moment when one wants the absolute equalization of temperature, pressure, etc.—and, therewith, the first two formulations of the Second Law—to be valid for an infinite amount of time.

Chapter 8
Translation from Dutch: Papers on the Pedagogy of Mathematics

Pauline van Wierst

8.1 Which Use Can Geometry Education Have for Students that Do not Continue with Mathematics?

The goal of this writing is to consider only that of geometry, which has common cultural worth, and therefore, justifies that it be taught also to people that do not have an aptitude for mathematics, and in their further life will not come in contact with mathematics, nor with its applications. I am of the opinion that such an investigation is highly pertinent in our days: not only weak students and their parents, but also mathematics teachers doubt the general usefulness of mathematics education. It seems as if the current curriculum and requirements for the final exams lack exactly that which could make mathematics—and geometry in particular—a subject for general growth. Or has it really just been an illusion, which motivated making geometry at schools with highly diverging objectives mandatory?

Geometry is concerned with Space. Nobody will deny that practical familiarity with spatial relations can be useful to anybody: for most things that we do is the capacity to perceive these fast and as thorough as possible of great value. The following, however, is often forgotten: the capacity to *see* and visualize space adequately plays often a very important role in *enjoying* the most diverse things in our world—either created by the hand of man, or not. One could really say that he, who is comfortable in dealing with spatial relations, distinguishes himself from

Tatiana Afanassjewa, *What can and should geometry education offer a non-mathematician?* (1924).
by
T. Ehrenfest-Afanassjeewa.
Translated by P. M. A. van Wierst.
Original title, *Wat kan en moet het Meetkunde-onderwijs aan een niet-wiskundige geven*, J. B. Wolters: Groningen en Den Haag (1924), 27 pp.

P. van Wierst (✉)
University of Amsterdam, Amsterdam, The Netherlands
e-mail: p.v.wierst@gmail.com

J. Uffink et al. (eds.), *The Legacy of Tatjana Afanassjewa*, Women in the History of Philosophy and Sciences 7, https://doi.org/10.1007/978-3-030-47971-8_8

somebody who is not, as a seeing person from a blind, and that spatial imagination should be developed from childhood on just as much as the musical ear and physical skills, etcetera.

It is, however, a different question, whether or not spatial relations should be brought to everybody's *awareness*, and if so, whether it should be presented as an ordered composition or as loose propositions. In short, whether exactly geometry education offers the most appropriate method to provide everybody with the knowledge of space.

Concerning the use of spatial insight *in action*, probably the same holds as what was already noted in other contexts: knowing—thinking—delays acting, dissolves it. A painter, a cox, a hunter, a cyclist, will use their spatial visualization the best, when at the crucial moment in their naturalness they are not hindered by knowing. Pythagoras' theorem or the calculation of the surface of a sphere would be of little use in such circumstances.

There are other activities for which knowledge of certain—especially also quantitative—spatial relations are very important. But on closer consideration, it turns out that also in this case the whole Euclidean framework is disposable.

Thus if it were just about the development of spatial visualization and practicing its application, then maybe in many cases other teachers could be more of help than the teachers of geometry. Geometry has, however, also another aspect which gives her particular cultural significance: the treatment of spatial relations has reached a particularly high level of logical rigor. Many expect therefore that acquaintance with geometry will have a special effect also on the students' capacity of thought.

That practice does not always live up to this expectation is known only too well. What, however, is the cause of this? Who wants to stick to his conviction (either pro or contra) can always do so: after all, one has every student only in one copy, and one can never know how he would have done on logic without education in geometry. Therefore, of course, everything that is said about this will be merely a guess. But fortunately, these guesses are based on experience which everybody can check for themselves, and therefore, these considerations don't need to be taken on trust, but should serve as an inducement for one's own opinion.

I am one of those that *do* believe in the effect of geometry for the education toward "being logical." I do think, however, that the instruction will have more success in this respect when one is clear on what "being logical" actually is!

Until the beginning of this century, one considered almost exclusively one side of the faculty of thinking: the formal-logical side. The failure that resulted from this emphasis for geometry education brought about a movement of teachers[1] that is characterized by the slogan "intuition." The representatives of this movement often even had an antagonistic attitude toward education in a logical direction.

In reality, however, what happened here was that the other side of the faculty of thinking, which is just as necessary to reach logic, and was instinctively brought to the foreground.

I would like to show that geometry education will only then be maximally fruitful—both with regard to the development of spatial imagination as well as of logic—when one attributes to intuition its rightful place in the process of thinking. Later more about this.

Here, I would like to say why I expect exactly from geometry this logical function.

It is no coincidence that spatial relations have been especially deeply logically analyzed. The same reason, which made it possible through the collaboration of many for centuries, makes it also possible for everybody individually to have this priceless experience: to organize a piece of his experience in a logical manner *by means of his own thinking*. The reason is the unpaired simplicity of spatial relations. No other subject,[2] with which human thinking occupies itself, is remotely as simple—unless one includes the rest of mathematics. With respect to the latter, however, geometry has the benefit that its material is familiar to every human being from its daily experience, and moreover is perceivably given.

If at least the faculty of thinking can indeed be *cultivated*, then that would be by its own activity, by judging, by formulating one's perceptions, and ordering them logically. One would be more motivated to do so when the material is presented intuitively, and is not dauntingly complicated.

In everyday life, one thinks little and only fragmented: one is usually content with separate judgments and does not attempt to unite these judgments into a consistent system. There are people who like this incoherence, a kind of vagueness in their views. Admitting such diverging views, still one should not overlook the following: the desire to express the experienced at least every now, and then and to be understood by others exists even within the most unreasonable people. One can see this in even the smallest disagreements: often embitterment arises caused by one's inability to express clearly what he intuitively sees, and the other's inability to take notice of only that, which is essential to the speaker. Practice in thinking and in expressing the insights that one acquired will surely do a favor to anybody at times.

8.2 What Is Being Logical?

Frequently, *logic* is contrasted to *intuition*: the words "logical," "scientific," and "abstract," are as it were used as synonyms in contradistinction with "intuitive" and "concrete." One holds that intuition is killed, annihilated by logic. In short, one believes that being logical means: turning away from intuition.

Against such a view protested already the most preeminent practitioners of the most abstract sciences, for example, H. Poincaré, and F. Klein, who also drew the far-reaching consequences for mathematical education. Also L. E. J. Brouwer highlights the role that intuition plays for the insight into the mathematical concepts.

Given, however, that views on logic and intuition diverge so much, the reason for this is probably partly also linguistic.

So far, analysis of the faculty of thinking did not advance so much that there exists a fixed terminology concerning it. The meanings that different people attribute to the word "intuition" do have something in common, but are not identical. I hope, therefore, that the reader will forgive me in case my usage of words differs from his.

My use of words is based on the following view: by acquiring insight there are always two steps which should be distinguished: *seeing* a certain feature in the picture that we have in mind and *becoming aware* of it. The element of "becoming aware" has a preeminent role in all the different steps in the process of thinking: by grasping and ordering that what initially is represented in our intuitive picture, by identifying the gaps and contradictions in it, by aiming to fill those gaps, and by tracing the origin of the inconsistencies. This all, I call the "logical" work. That, which in this manner is processed (or sometimes also remains unprocessed), I call "intuition." Disclosing something without being aware of it and also its unconscious ordering I count as the "intuitive" work.

That such a division between the conscious and the unconscious in the procedure of acquiring "insight" is, in any case, justified, and can be shown with the example of unconscious action, in which case there is undoubtedly some form of perceiving, and in which case there cannot be any *thinking*—of *logic*—(for example, a quick movement to avoid a threatening danger, before one understands what had to be done and why). However, what seems to me to be very essential and what, as far as I know, normally is not pointed out, is that *logical action without intuition is impossible.*

One likes to tell that Gauss said about one of his discoveries: "I already found the theorem, I just did not prove it yet"—which is used as a proof that also in mathematics, in an intuitive manner—without "logical thinking"—one can gain insight. Also H. Poincaré talks about how searching and finding mathematical facts often happens in an unconscious manner, and he concludes from this, how very necessary intuition is for a mathematician. These are, however, all cases in which the disclosure and the awareness are very clearly separated from each other in time.[3] I want to stress, however, that also in cases where understanding happens in one single—timely unseparated—moment, the two movements are present: always also the intuitive. *Without intuition no thinking is possible.*

It is often said: "he made a mistake because he relied on his intuition; he should have used his logic."—in our way of speaking, one should say: "he made a mistake, because he used his *intuition in a sloppy way*: he should have ordered the first impressions that his intuition gave him better, *that is, treated them logically*: then he would have discovered that not all these impressions have a place in the *complete* intuitive picture, that something or other was in contradiction with the whole and should have been replaced by a better (intuitive!) element."

But: is this all relevant for *scientific* thinking? Isn't scientific thinking a very special process in which intuition is replaced by logic? After all, we did not speak about *formal logic* yet!

An old saying goes that one can explain clearly, what one understood well. That is why a proof which is formulated in syllogisms is a *sign* that one analyzed the material adequately deeply.

It is, however, not correct that the syllogisms are the *instrument* itself of thinking. Does looking for an answer consist in the concatenation of syllogisms? Can somebody else, that we show our syllogistic argument, follow us, without *thinking* himself … in a non-syllogistic manner?[4]

The logical form—not only the syllogisms, but any somewhat purely one-dimensional concatenation of thoughts—comes forward at most as the closure of the thinking-process or as starting point for new research, because, as mentioned before, one can see from this that the question has been analyzed adequately deeply. But even though this kind of formal processing is used frequently in the process of thinking, it is not the case that they are the thinking!

Understanding the object does not happen in those moments in which one abstracts from it and considers the formal-logical relations between the propositions that describe it: in such moments, of course, one thinks, but not about the object but about something else—about these formal relations. One relates oneself to the object not as "thinking," but as "calculating." For example, the algebraic treatment of the formulas in physics is not thinking about the physical relations themselves, which are represented in these formulas.

That I reserve the term "thinking" for this one thing only: for the treatment of the intuitive material through consciousness, might be a matter of terminology. But the essence of the matter is, that in scientific research I distinguish between *two* things: the "thinking" and the "calculating"—in the above sense—and that therefore I do not refer to both with the same term "thinking."

There *are* things that we discovered by means of calculating *only* ("calculating" now in the usual sense). But anyone will acknowledge, that he possesses knowledge obtained in this manner in a completely different way from knowledge obtained by "thinking" in the sense that I attributed to it. Knowledge obtained by calculation can easily be wrong and one can (unless one also tried to give her a place in the original intuitive picture by means of "thinking") be fooled: many a good scientist can tell such a story from his own experience.

But the same also holds for the application of formal logic. Moreover, one uses her rarely to come to know something new. Also when one is formulating proofs, the real difficulty is to *come to know and bring to light the premises, which are essential to the case*: the formal-logical reasoning itself costs usually no effort— and as such gives no new insight—it is an automatic consequence, as it were.

About a problem can be difficult: becoming aware of what one is searching for, how one should shape the question so that the subject is presented in the most effective way—because also this initially is seen intuitively and sometimes only much later and with a lot of effort brought to consciousness; subsequently, the ordering of the material which is given by intuition, collecting everything which is relevant, and discarding everything irrelevant. Who can do *this* is *logical.*[5]

The actual work of logic happens at those moments in which intuition is brought to consciousness.

To illustrate what I mean with "logic" and "intuition," and for later use, I would like to make the following remark. In geometry ever since Euclid, two completely different branches of science have been conflated: the *Study of space* and the *Axiomatics of geometry*. The intuitive material of the study of space is that, which is given by *spatial imagination* (it is here not important, from where the latter comes). The logical elaboration of it exists in finding the most essential spatial relations, formulating and identifying them; also in identifying relations, which go beyond immediate imagination, but which we nevertheless do not doubt, because we see a connection with the spatial relations familiar to us.[6] It makes that our knowledge of space becomes richer and clearer.

The intuitive material in axiomatics are all the *axioms*, which comprise for us the study of space[7]; the logical elaboration of them consists in bringing to stage the most essential of those, from which the others can be deduced logically, and showing their (formal-)logical independence; she yields a more perfect system of axioms, one in which the logical coherence has become clearer.[8]

The content of intuition can consist in sensory perceptions (like in study of space), but also in the results of a former logical elaboration (like in axiomatics). This is the contradistinction between "concrete" and "abstract."

When I elaborate a certain intuitive picture logically, then I focus on its particular characteristics, so that all the others are driven to the background: this is "abstracting." When in this picture I am especially interested in one characteristic, which I do not understand at first, then it is very important, that I pick out only those elements of the picture, which have something to do with that: to be able to *think* I must master the art of *abstraction*. In this sense, one can admit that "being logical" coincides with "being abstract." But one has to keep in mind that in order to be able to be abstract, the intuition from which to abstract, is unconditionally needed.

When one expresses the result of thinking in words, one gets abstract sentences, that is to say, such that display only a few elements of the intuitive picture. But he, who formed them, possesses more than this. And this holds in general: when somebody states that the expressions of the other are "too abstract," then he himself is the one which misses the necessary intuitions, not the other which is blamed for his abstractness. The other is of course not excused by this: when he wants to communicate his insights, then it is *his* task to make sure that everybody gets the same intuitive picture, from which he himself reads his judgments and proofs! Nevertheless, this is the work of an artist and for a normal human being in most cases extremely difficult. But in many cases, undoubtedly many difficulties could be avoided, if one did not assume the false axiom that all people have possessed from the beginning of the same intuitive picture.

The ultimate goal of thinking—which often is reached only partially[9]—is obtaining an *intuitive* picture, that is more perfect than the initial one and that one comprehends well. In this sense, it is thus not true that logic kills intuition, as is often sustained; to the contrary, she enriches her.[10] I hope that I managed, with my terminology, to distinguish these two elements which can be developed in education by very different means.

8.3 Can Geometry Education Promote the Development of Logic?

In connection to all the foregoing, my answer would be yes, as long as one does not fail to create those preconditions without which thinking itself is impossible. These preconditions are a sufficient amount of intuition and the curiosity to analyze it. In general, little or nothing is done to this aim: some of the teachers do not know how necessary it is to take care of this for many students, they believe, that the comprehension of spatial relations is formed synthetically with the help of the axioms of geometry (that these axioms could be something else than the excerpt which came into existence by analyzing the intuition which had already previously been given); others presuppose that every student already possesses from itself the necessary intuition—and then build upon something, which in many cases does not exist (sometimes one sees a remarkable phenomenon: spatial imagination is present, but because of a misunderstanding it is not taken into account in the geometry class).

But even if one had only students that already beforehand were capable of good spatial imagination, then still could several factors in the current common education diminish their receptivity for "logical rigor":

1. In the first place, it would be the mix of two sciences which was mentioned in the previous section: the study of space and axiomatics, whereby two different conceptions come into play, without them being clearly distinguished from one another. At the same time has the alarming word "proof" in these two sciences has two different meanings. In the study of space one understands "proving" as *giving insight* into the *correctness* of the proposition. In axiomatics it means: *tracing* the proposition *logically back to the axioms*. In this sense, no one would object to accept a proposition as "proven," even if no one believes that she is valid in empirical space—if one is considering an axiom system which does not hold for empirical space. In this last meaning, on the other hand, a proof is required, even if there is no way to make the students doubt a certain proposition—as long as this proposition is not an independent axiom.
   If the teacher is not conscious of this distinction, and, as a faithful servant of science, tries to convince his students that without the proof of an obvious proposition, their whole geometrical knowledge is built on uncertain foundations, then this can easily make them doubt the "seriousness of science."
   Of course, it is no coincidence that views stemming from both sciences are intertwined in education: they do really turn into each other unnoticeably and different individuals will draw the line at different places. This is a stronger one's capacity to think; the more propositions will not be obvious to him—for he will be more capable of indicating different possibilities.
   Let us consider, for example, the question: draw a circle through three points. To find the center, as is well known, we use only *two* of the three connecting lines. The center is the point where the line segment bisectors of the two lines cross. The proof does not give difficulties. But then the question arises: will the line segment bisector of the third connecting line go through that same point?

Intuition will expect this immediately. Somebody, who is logically inclined, will nevertheless have the urge to identify the reason for this, which he will do easily. But for a mediocre student, it will be hardly possible to focus his attention on this. Nevertheless, the question might become more meaningful to him, if one formulates it like this: "doesn't one obtain three different circles, when one starts every time with a different point?" (Think about the construction of a triangle, when three sides are given: one gets more than one triangle, even though intuition seems to say initially that the triangle is determined *completely* by the three sides!)—Often, in this manner, the goal of a proof which seemed axiomatic can become clear.

I am far from the opinion that the axiomatic point of view should be eliminated from education. To the contrary, I think it has great *practical* value, that somebody is interested not only in the correctness of his views, but also in their origin and logical reasons; this is also a better guarantee for the correctness of the views, or it helps to see its relativity: much bigotry could be avoided, and many new possibilities could be realized in this manner… But also in any normal occasion, in which we want to test a certain proposition, we must draw conclusions from uncertain premises and only later on the basis of the conclusions decide, whether or not the premises are credible—and that is called adopting the axiomatic point of view!

But how to make students familiar with the axiomatic way of thinking, that is another question. Often it is said: the infantile mind is not receptive for "scientific thinking" and requires explanation which is more directed at intuition.Now, how necessary intuition is for all thinking and all ages, we already discussed. The indifference, however, that many students show after the first geometry lessons for the "logically rigorous" proofs, can very naturally be explained, by that it is in itself impossible to be interested in the viewpoint of axiomatics, before one got to know the system of propositions that have to be axiomatized, that is to say, the study of space. Relatively few individuals are instinctively inclined to trace back one proposition to another, without questioning the goal of it. Most of them can only show interest in the study of space, which they take—really not unjustly—geometry to be. And with this, they get enough opportunity to sharpen their logic. And if one would like to give way to this natural inclination, this would at the same time be *more logical.*[11]

The introduction to axiomatics and also to questions of an epistemological nature (of course, not to an exaggerated extent) would be much more in place after the systematic course about the study of space. If the teacher would then manage, then this would, in my opinion, contribute more to the development of the student, than extending his knowledge with a couple of artificial moves to solve problems. I believe by the way, that this also holds for students which will later specialize in mathematical subjects.

2. The other danger lies in the common method of education, that maybe one can justly call the "method of insisting." It consists in trying to get the students to imprint the results of the thinking of others, by letting them repeat proofs which are handed to them completely, expecting that in the long run they will

get so much used to the form of proof, that they will start to think more clearly themselves. Hereby, however, one forgets the following: the Euclidean form of proofs is a meaningful summary of the *results* of thinking, but not a *reproduction of the process of thinking itself*. And therefore one cannot learn from it how to search, but only how to formulate the findings.

To use this benefit, one should first search and find the thing oneself; only then one can appreciate, to which extent a pure and short formulation will contribute to completely remove the last unclarities that might have remained in our thoughts.[12]

There are, of course, students that understand the issue immediately. For them, the Euclidean form of explanation is a sufficient *hint* for what they should imagine, in which way they should think *themselves*. They do not rail with "abstract," because they already possess the necessary intuition themselves.

But for others, one should first make sure that the pertinent questions do not appear abstract, that they accept the goal of proving: *then* they will also start to think as well!

3. A third danger for logic is overloading the course. Euclid showed two precious things: (1) how one can reduce a proof to the essential, and (2) how one identifies the most important in intuitive material and organizes it in such a way that every element of it can be acknowledged and proved with the greatest possible ease.

   He who absorbed the spirit that is contained in here, from him it can be said that he has learned from Euclid to be more logical. This spirit is, however, forced to the background when one includes propositions in the buildup of the proof which are not strictly necessary for the deduction of those relations, for which the whole buildup is taken on. That, which in elementary geometry is the final result, are the quantitative relations for the calculation of the area and volume of the sphere. That, what one necessarily needs to know to deduce *those*, suffices also in any other area that has to do with spatial relationships, at least as long as one does not want to go into particular depth. They are the propositions about congruence and similarity of shape of triangles, Pythagoras' theorem, the calculation of the surface of parallelogram and triangle when basis and height are given; the calculation of the volume of a parallelepiped, the theorem concerning the equality of volume of pyramids with equal bases and heights, and finally the concept of limit in the definition of quantities that come up in the treatment of the circle, the cone, and the sphere. For someone who in the future will never be concerned with geometry anymore this definitely suffices.[13] But when he needs to learn many deduced propositions and applications, which of course are less simple than the fundamental system and therefore will preoccupy him more, then he will lose the thread and will remember, after the end of his education, only some particular disconnected theorems.

   Many teachers are of the opinion that a theorem from the fundamental system (let us take for example Pythagoras' theorem), is only understood well, when students have practiced in applying the theorem to various questions. But when one has a closer look at the matter, one notices that for such applications the difficulty is not the application of Pythagoras' theorem, but rather the use of various artificial moves, that require not the understanding of the theorem, but

some special geometric innovative power, which one cannot cultivate in non-geometers anyway. Whether the theorem will be understood or not, depends, however, on what the student had learned to master beforehand. How *important* the theorem is, is best displayed when one shows that she is the ground for all important calculations that are discussed in the course, and also how often it can be applied in practical questions (without artificial moves!).
Learning many deduced propositions, which all in one case or another can be used as "auxiliary proposition," has another damaging consequence: the student tries—when it comes down to it—to bring out one such proved formula which has been useful at some point from his *memory*, instead of trying in a natural manner, by application of just a few fundamental propositions, to *work through the problem himself*. This applies most of all to the projection theorem, Stewart, etc., which anyway are nothing else than modified applications of Pythagoras.
The purpose of logic is most of all to, as it were, concentrate that what is investigated, to accomplish the connection of the whole area using a *small* amount of strings. Therefore, it seems to me that an *overload* is simply an *offense against the requirements of logic*.
Just as Euclid's proof method is appreciated only then when one by investigating it made it one's own, in this manner also can the conformation of the whole system become clear to somebody just through his own work. For this, the opportunity can be given by showing right at the start the goal of the investigation, and also the most prominent steps that have to be taken to reach it. The students can then try to establish what one needs to know to answer such an intermediate question and what in turn is needed to know that. Those, who are unable to do this, will in any case, look at each proposition that should be proven with more understanding.
The students will take an active-penetrative attitude with regard to the system, when one does not immediately start with the systematic course, but instead precedes this with a series of discussions and exercises that give the students a much broader insight in spatial relations as a foundation. The systematic course then presents itself as a purposeful analysis of everything that one has seen in "real" space, as a solution for questions that one encounters in practice. These questions will regard, however, mostly the last propositions of the course—and in this manner will at the same time the whole course be given a direction.

The program, of which I would expect success for the development of logic, would thus be the following:

1. In the first place an *introductory course*, in which no theorems will be proved, but preparing exercises will be made to develop spatial imagination. About this more in the next chapter.
2. A *systematic course*, which, however, should be different from the common one with respect to the following:
(a) A proposition will only then be proved when there is at least one student in the class for which it is not evident; evident propositions will explicitly be accepted as (temporary) axioms. (b) Determining, formulating, and proving propositions will happen—not without guidance of the teacher, but still with far going input

of the students themselves. After an "introductory" course this will be possible. (c) The content of the course should be as concise as possible.

3. Axiomatic revision of that which has been learned, in which many of the "temporary axioms" will be "proved."

After that the students, by means of the problems of the systematic course—which they always have been able to accept with their own rcasoning—practiced with proving techniques, also this will be easily doable.

Some introduction to the questions of axiomatics and epistemology would be welcome, but should not be obligatory for every teacher and every class. The whole education should be impregnated with the connection between the geometrical propositions and the spatial relations that one encounters in the material world. The students should be able to recognize the concepts and relationships that they learned in geometry in reality.

8.4 The Introductory Geometry Course

The intuitive content of the study of space will be acquired by sensory experiences: one can have different views concerning the aprioricity of spatial representations—nevertheless one can easily experience that those configurations which one has seen and felt more are easier to imagine. For this reason, it is definitely an improvement, that the geometry teachers now will direct their attention toward the "laboratory method" of education. Nevertheless, I do have something against textbooks written in this spirit. What I have seen in this area (it could be, that a book which is different, escaped me!) has a common inclination: what one finds is more an escape from logical rigor, than an enrichment of the course with spatial images.

It comes down to proving Euclid's axioms anyway, only one does not prove them "more geometrico" but "more physico"—or something like that.

This shows, however, that one also misunderstands the physical method: nobody will create a physical experiment in order to prove the consequences of well-known facts: those will be deduced in physics, just like in geometry, by thinking—for this one has his ability to reason, to save himself from the redundant experimenting!

For the questions in Elementary Geometry at school there is no reason to bring about experiments: the fundamental hypotheses are also like this, immediately clear for everybody, as soon as he only understood what they are about; deviations from this in the material world—which are made probable by the theory of gravitation—can anyway not be detected with the accuracy of student-experiments; whereas the consequences of that, with which one is occupied in the normal course, do not introduce such difficulties calculation-wise, that they justify substituting experiments for calculations.

Moreover, experimenting is not always the best way to provide for the necessary intuition. Compare, for example, the proof of the proposition about the sum of the angles in a triangle according to the experimental method with its proof according to the method of Euclid. A purely empirical proof would be the following: one measures the angles of various material triangles—for example, in degrees—and every time one sums up the three numbers thusly obtained.

Summing up these numbers, one does not experience that what one is summing up are the three angles of a single triangle; this circumstance has here the pale role of *abstract* knowledge. Euclid, to the contrary, lets one draw a parallel line, with the help of which in thought he conjoins all angles together at a common angle point. He, who earlier has *seen* the equality of angles in case of parallel lines, will *see* now not only, that these constructed angles together form a right angle, but he will also see that it is *because* both the constructed angles lie on a same straight line (the basis of the triangle), which is parallel to the auxiliary line, and *because* both transversal lines (the sides of the triangle) intersect in the third angle, in short: *because* they are angles of one and the same triangle.

It is, of course, not difficult to introduce also in the empirical treatment of this problem those elements that Euclid pointed out and that promote intuition so much (by, for example, not expressing angles in numbers, but laying them out as geometrical quantities—for example, to cut two angles of a paper triangle and hold them against the third angle). But that, which I wished to stress, is that many a logical method entertains more contact with intuition than most experimental methods.

Often one loses sight of the fact that exactly the Euclidean[14] ordering of the material is so fit for this.[15]

The accomplishment of logic by Euclid is exactly this: ordering the visually given material in such a way that the theorems that one needs come to hand almost by themselves.

While the proof method can do nothing else than simply establishing certain rules *without connection to the others,* Euclid makes that one *intuitively sees* them, by putting them in a visual connection with others that one already has acquired insight in. The logical method thus has something beneficial concerning the certainty and the universal validity of her solutions, and also concerning her visualness.

One can still raise several objections against the experimental treatment of Euclid's axioms.

If one plans to treat them later in the systematic course anyway, then one should not already beforehand take away from the students the joy of discovery (by means of thinking).

Further: if one first declared the empirical confirmation of a relationship a "proof," it will not be easy to move the students to listen to the proof again. And, furthermore,: the total content of a geometry course is not of such a nature that students would have the same interest in it if it were repeated. After all, the propositions treated in it are not so much interesting in themselves, but more in virtue of the relationship with a broader and versatile experience which they account for, and for which they are a manner to have a complete overview of it from one common viewpoint.

Giving students this experience must, according to my opinion, *be the goal of the introductory course*, and the laboratory method should not make them familiar with geometrical propositions, but with *geometrical concepts.*

I want to make this clear again by means of the example of the sum of the angles in a triangle. Even though I think it is very wrong to teach the general proposition concerning it in an empirical manner, I do not think it is very bad to let the students on this occasion first estimate the sum of the angles of some paper triangles, and then measure it. One does not consider enough for how many students the words "angle," "summing angles," "triangle," "measuring," etc. are "abstract"! The execution of empirical manipulations makes that students fix the meanings of those words much more lively in their minds; the repetition in altered special cases makes those concepts broader. The goal of such exercises must be that by mentioning a geometrical figure, the student can imagine every special case, spatially oriented in every way. The drawings on the blackboard should for him not be the *objects themselves* of the education, but only the schematic representation of that which he has seen and imagined, and must be symbols of the concepts which he constructed from it himself.

There are so many questions, for which it just comes down to imagining the configurations well, for which no background knowledge is required, and which are even for the best students not trivial, but that would find no place in the systematic course. One can connect them with problems of a practical nature (with "practical" I do not mean merely "useful in the battle for existence," the problems can also connect to the interest in games or to the aesthetic or scientific interests of the students). With that they learn that space is something about the physical world that merits special study. After all this, a systematic research thereof would present itself as something natural and desirable. Familiarity with the most important geometric concepts and notations required beforehand would avoid an evil that often is experienced at the beginning of a geometry course: one has to battle different difficulties: not being used to the terminology, as well as the lack of imagination, so that the last difficulty, which actually in the systematic course requires full attention; the logical elaboration of the material will get insufficient attention.

The teachers that interact with students that have had an introductory course conclude how much they profited from it.[16]

Considering the nature of the exercises—the most beneficial would be, as said before, the laboratory method, that is, working with material things consisting in measuring, drawing, sticking, sculpturing, etc. Since, however, the goal is developing the imagination one should aim to obtain, which the students first try to imagine, every figure under consideration and then only afterward test and correct their imagination to match real objects.

How long one should continue such exercises would depend on the general development of the class—not longer than is required to prepare the students for the systematic course. And in any case, one should aim for them starting to guess for a certain regularity in the existence of spatial relations, so that in the end they will desire to establish these relations exactly—or at least find it very natural to occupy themselves with that.

8.5 Some Examples of Questions that Would Be Suitable for an Introductory Course

It seems wrong to me, to write down a systematically ordered collection of exercises for the introductory course that I have in mind; it is not about things that one should "learn" at all schools in the same manner. It is better when every teacher according to his own taste and in accordance with other tasks and the character of the students selects the examples and gets as much as possible out of it.

However, I heard from many teachers that the subject of geometry is so limited that one is forced toward the Euclidean axioms, if one searches something that can be understood by beginners. Therefore, I would like to mention some categories of questions, just to show how large the choice is in reality.

1. *The difference between the right and other lines.* Examining whether a ruler is straight, by putting it in different manners through two points on a sheet of paper and drawing a line alongside it. Criticizing the method (if the ruler is bended, then this cannot be discovered by a movement alongside those two fixed points; turning about a fixed point *in* the surface of the paper sheet is not decisive either: which divergence of straightness escapes control in this manner?) What happens to the edge of the ruler when it is turned around its axes through two fixed points?—if the edge is straight and when it is not?
   The way that is taken by the various points of an object, when one fixes two of those.
   The axes. Describing the rotation-surfaces by means of different other lines. Searching rotation-surfaces in the room and fixing the shape of the lines, that can be rotated in order to describe them. Application by working with the lathe, the fabrication of pottery.
2. *Continuing the straight line in the infinite.* Imagining clearly the straight line that goes through two points within the room. Where does she penetrate the wall or the ceiling and the floor of the room? How does she continue through the rooms upstairs or next door?and further? Elucidate by means of boxes that are connected like the rooms are. How is the line positioned with respect to the rotation axes of the Earth? Explain using the globe.
3. *The straight line as the shortest distance between two points in space.* A drawn cord between two points, when there is nothing in between and when there is an object between them—imagine the form of the cord in both cases. A drawn cord on a surface. The shortest distance between two places on Earth—measured through the globe or on the surface. In which direction goes the shortest line of connection (*on* the surface of the Earth) between Rotterdam and Batavia (South, East, North, West)? Establish on the globe! Geodetic lines on different surfaces.
4. *The light ray as a straight line.* If I want to look at a point through two holes, how must those holes be positioned? What should change if I put in between the holes a glass prism?
   The form of the shadow of an object. The space of the shadow, its borders—the cone.

5. *Estimating* and then *measuring of lengths* of different objects in the room. Estimating how thick a circular cylinder will be, that one makes from a square piece of paper.
6. *The difference between the plane and other surfaces.* Can I fold every surface? Which surfaces can I fold? Can I fold a cone on any of its points in any direction? Which surfaces can I unwind on the table? Constructing ruled surfaces from paper and drawn cords.
7. *Position of straight lines with respect to each other.* Searching parallel and intersecting lines within the room. The position of the different edges of a tetrahedron and other polyhedrons with respect to each other. Searching for a definition of parallel lines.
8. *Angles.* Comparing the size of different angles of polygons that constitute the junctures of different objects in the room: estimating and measuring by making paper angles. Sum of the angles that come together in a vertex of a polyhedron. The angle on the top of an unwound cone surface. The angle, which the minute hand of a clock in a minute, a quarter on an hour, an hour, three hours describes. The angles under which we see different distances. Apparent decrease in size of an object when increasing its distance from us. Estimating the angles, under which we see different objects, and checking the answers by measuring. Estimating the angle under which we see the moon, in comparison with the angle under which we see other objects, checking by putting ourselves in such a way, that those objects cover the moon (an interesting experiment, that shows how we err in estimating the size of the *angle* of the moon).
   Central angles and arcs. Straight angles. Trying to give a definition of those. Creating them by folding paper.
   Around which angle did we turn on the entire route through all those streets from our house to school?
   Angles between two flat surfaces. Three- and more-way corners. Viewing, imagining, and drawing after memory different polyhedrons.
9. *Rotation speed.* Comparing the rotation speed of the three hands of a watch. Rotation speed of trains wheels of different sizes, of the Earth. The relative speed with which we see from the train objects move which find themselves on different distances from us. The moon walks with us. Explain this.
10. *Symmetry.* Central symmetry on the plane and in space. Symmetry axes of second, third, etc. order. Symmetry line, symmetry surface. Looking in the environment for examples and creating them themselves. Models of crystals.
    By the creation of the simplest models from card board arise spontaneously the more quantitative questions, which make the transition to the systematic course very natural.
11. *Cross sections.* If I insert a plane through a point of a dihedral angle, then the cross section will be an angle. For which position of that surface do I get the biggest, and for which the smallest angle? If I let this surface go through a straight line that cuts the dihedral angle, then still it can take infinitely many positions. How does the angle of the cross section change, when I let the surface turn around that line?

Cross sections of polyhedrons—imagining, drawing free-handed, testing with plastilin models. Change of the cross section by parallel movement of the cut, by turning it around a certain line.
Shadows on a wall. The sun as light source: can the shadow of a thin tense wire be four meters wide? Which shape can the shadow of a sphere take? A "pointy" light from a lantern as a light source: which shapes can the shadow of a coin take? Put a cover on the lantern with a circular opening and study the shape of the bright spot on the wall at different orientations of the lantern.

12. *Cutting figures from cardboard*, from which one can make certain models by folding them—without having to cut them in pieces. The pattern of a cone, of a skirt, of a light shade, of a cube, of the model of a house.
13. *Degrees of freedom.* How much data is needed to determine the position of a point in space, on a surface, on a line? Which data can that be? I know that somebody is on a distance of 1000 meters from his front door, where should I search for him? How many and which data are needed to determine his exact location? How many different triangles can I draw from a given basis and adjacent angle? And when only the basis is given? With which points on the surface can the top coincide—in the first, and in the second case? How many degrees of freedom does a certain mechanism have (for example, a bike, a sewing machine, etc.)? How many angles can I draw with three things given? When three angles are given? The latter is also a question which can be used to introduce the systematic course.
14. *The topology of lines in space.* Something which for many people seems to be much harder than the topology of surfaces. Using the stereoscope can in this case be very helpful.
Drawing a knot clearly—first from fantasy, then checking. The topology of the thread in a short piece of knitting. Different kinds of knots. The Möbius band. Puzzles. The cross section of two cylinders, of two arbitrary second degree surfaces.
The degree of difficulty that these exercises raise for the imagination is, of course, varying. How far one should go in considering every single configuration will depend on the capability and the talent of the students, but also on the courage of the teacher and his ability to find suitable objects with which the matter can be demonstrated.
To him, which is skeptical about such merely qualitative treatment of geometrical configurations, I can say the following: in every systematic and exact treatment of the subjects mentioned here, one should already possess the imagination and the qualitative estimate beforehand, in order not to be desperately helpless in confrontation with it: this cannot be built synthetically, on the basis of the correctly proven propositions of the systematic course; to the contrary: these propositions speak only to those, who have already before the study of geometry taken in enough spatial images. That, which "good" students without help of the teacher bring along and with which they surprise others, that can to a considerable extent also be taught to many other students.

Notes

(1) Especially in Germany, America, and Russia.

(2) Just compare an arbitrary polyhedron and everything that one discovers about it with a horse, a flower, a dish, a sonata, a relationship between two people, an historical event…

(3) I mean: disclosing those relations upon which the theorem meant by Gauss was founded and that necessarily must have been present in his intuition, if he were so sure of his theorem—the theorem was after all itself *logically* clear in the mind; or also disclosing that intuitively given area, where the fantasy of H. Poincaré was roaming, before he managed to capture that which he searched also with his consciousness.

(4) Have you ever mastered a mathematical proposition *really* by merely checking a rigorous proof in a book step by step, without having intuitively apprehended the relevant subject matter as a whole?

(5) Somebody, who has from the start a clear intuition concerning a certain subject can give purely logical, that is, harmonizing, non-contradictory utterances, without being really "logical"—in the sense given above. The lack of logic will show in his case as soon as he finds himself on terrain where he can acquire the insight only by means of searching it *consciously*.

(6) One sees for example easily, that between two sides and an angle of a triangle, and its third side, there must exist a relation. That in case of a straight angle exactly this relation is expressed by the Pythagorean theorem probably nobody will see immediately: but they wíll see the perceptible connection at every step of the proof.

(7) Also the relations of subsumption between the given propositions and the formal-logical structure of every proposition are *seen* before one can *formulate* them *consciously*: also the logical relations are first understood *intuitively*!

(8) The construction of this task of science would in principle not change, if it were propositions about arbitrary things and not exactly about spatial relations. Now this is actually not completely correct: in closing the gaps in *this* intuitive material, one uses again and again spatial imagination. Besides, axiomatics had to serve the epistemology of space all the time, and in the beginning was not even distinguished from it.

(9) I mean with this that many results remain to us only temporarily *proven*, and not *intuitively understood*.

(10) That which really can be killed or numbed by too much consciousness are the *intuitive responses*. These are in many cases highly desired and should be spared by the educators! By the way, *most of the time* consciousness causes confusion only at its first occurrence together with an instinctive action. After one got accustomed to it, the instincts will go ahead again. But maybe sometimes it takes way too long before one gets this far!

(11) One should thus renounce proving in the beginning propositions which are evident for all students in the class. But one should not assume them implicitly.

(12) Sometimes one tries to remedy the lack of the own activity of students by letting them prove, in the form of problems, some propositions which are not needed for the conformation of the foundational system. These propositions are, however, often more complicated than those of the system itself and can oftentimes only discourage the students: proving them requires often not so much thinking (that is, analyzing the intuition) but rather inventing (of special auxiliary structures), which thus requires a special talent, which has nothing to do with our problem: the logical development of the student.

(13) And for the further development of specialized mathematicians, it seems to me that the eternal combination of ever again the same concepts from Elementary Geometry—which all extensions and applications that the normal course contains come down to anyway—are of no use whatsoever. A mathematically inclined student would benefit much more from using his study-time to open up a broader horizon of mathematical research.

(14) When I talk about "Euclidean" ordering, then I mean with that every ordering, which preserves the spirit of Euclid and which possibly articulates him even better than Euclid himself did.

(15) For the study of space is contact with intuition which originated from sensory perceptions the best; for axiomatics, on the contrary, it is not suitable at all. But therefore one cannot sustain, as some do, that using imagination in *geometry* per se is "unscientific."

(16) I did, however, hear people say as well that the "introductory" knowledge of the *propositions* in turn obstructs the willingness to accept "logical rigor"—and this is indeed in agreement with what I said above about the "laboratory proofs."

(17) Which "angles" where intended here?